Optik and Image

How small can we see?

A brief summary of light-image relationship

By

Dr. Kartika Padhan, PhD

Acknowledgement

I was born in a remote village in the state of Sun Temple, Odisha in the eastern part of tropical India. Growing up in a farmer's family, most of my childhood was spent in the sunny farmlands along with my father. Since then I have been fascinated with the beauty of nature. A farmer's life is very closely associated with nature. As a child, so much joy spread in the face whenever I saw a rainbow in the sky while curiously thinking that the ends of rainbow must be touching somewhere on the ground. Many times, staring at the mystery of the Sun radiating brightness upon the world, I ask myself what exactly the sunlight is. Sometimes I wonder how we see the world. Even more, how do other animals see? When I see an eagle from far above the sky swooping down on a mouse in the blink of an eye I wonder how it can see from so far. On the other side when I see an owl catching a mouse at night I wonder how an animal can see in the darkness. These curiosities have been the source of inspiration for me to write this book. Thanks to my father (who passed away when I was 20), mother, brothers, and sisters for giving me the company of natural childhood.

As microscope is the most important tool of my research I have long thought of writing a book on what light is and how it helps us to see through the big eye- "Microscope". Many times, I struggled to understand the basics of image formation. To understand the relationship between light and image in microscopy, I did many simple experiments at home with the help of my young son Abhilash. Without his curiosity, these experiments wouldn't have been possible. The burden of family responsibility and academic workload makes it difficult to find time to think of science un-related to the job. Thanks to my wife for all her support despite taking complete care of my son and daughter.

Throughout my life, I have been lucky to get very good teachers. Thanks to all of my teachers for helping me learn and grow as a researcher. Especially the contributions of Mr. Prafulla Kumar Pattanaik, Mr. Artabandhu Mishra, Dr. Shahid Jameel and Dr. Rajat Varma are priceless. Nonetheless, I owe all thanks to you- the readers of this book!!!

Kartika Padhan
Rockville, MD, USA

Contents

I	Acknowledgement………………………………………..2
II	Introduction……………………………………………….5
II	Vision ……………………………………………………6
III	Image …………………………………………………10
IV	Light …………………………………………………12
V	Wave-particle duality………………………………….17
VI	Pinhole camera……………………………………...…36
VII	Lens …………………………………………………40
VIII	Microscope ………………………………………….....45
IX	How small can we see?..69
X	Electronics in the eye………………..…………………83
XI	Conclusion……………………………………………....88
XII	Glossary……………………………………………….90
XIII	Index…………………………………………………..93

Introduction

"Light"??? Time and again I wonder what exactly light is. Why do the oceans and sky look blue? Why does the sun look bigger during sunrise and sunset? Why do stars twinkle? What do light's components look like? Is light a wave or a particle? Why do we see seven colors of light? Answers to these questions can be well understood if we analyze the basic properties of light.

Through the ages, human beings have always been fascinated by light. Light is the most visible thing that we see and that helps us to see but there is nothing that helps us to see light. Discoveries of scientists have made it possible to understand several properties of light. Even if we do not know the exact nature of light, different properties of light can explain different phenomena associated with it.

Many times, we look for wonders far away in the universe, but there are no fewer wonders within our own body. Light enables us to see very far (as in the telescope). It also enables us to see very small things (as in the microscope). My curiosity is to know what is the smallest thing that we can see with the help of light. Here in this book, I will discuss different aspects of light that help in image formation from a simple pinhole camera to super-resolution microscopes with the illustration of some simple experiments done in my own hands. I will start the book with vision and how light helps us to see. Then I will pursue to explain the very basic properties of light. Eventually, I will try to explain how small we can see beyond the power of the human eye and what the limitation is beyond that.

"He who sees with his eyes is blind".

-Socrates

Vision

I will start with an illustration about the perception of vision from ancient Greek philosophical work "Allegory of the Cave". Socrates imagined some slaves locked from childhood in a cave with their neck and legs tied to the wall. There was a fire behind the wall, which produced shadows of passers-by carrying puppets. The slaves could only see the shadows on the opposite wall. Having never seen the real passers-by, but only hearing the sounds of their talks, they believed that the shadows produced sound. When one slave was freed to see the reality, the brightness of fire blinded his eyes and thus he could not see the real passers-by, which further strengthened his belief that shadows produced the sound. When he finally came out of the cave to the outside, the overwhelming sunlight blinded his eyes initially, but gradually he adjusted to the sunlight and gained vision that abled him to see the real world. Finally, when he returned to the cave, the darkness of the cave blinded his eyes again. The locked slaves thought that the outside world damaged his eyes and instead of believing the truth they continued to believe that the shadows were the reality. Socrates' saying that people who look through the eyes are blind can be well understood from the interpretation of this allegory. I will explain the scientific basis of this allegory in the last chapter of the book.

"Seeing is believing". We always believe in what we see, even though that may not always be true. Probably that's the reason why ancient people believed that the sun orbits around the earth. Although we have five senses- hearing, sight, smell, taste and touch- our brain prefers to believe sight the most. If we see something and hear something else, we will first believe in what we see. As an example, if I just move my fingers on the piano to the tunes of an already recorded song, the viewer will believe that I am producing the music on piano even if the sound comes from a music player. One reason why sight is the most powerful sense could be that the visual cortex (the part of the brain that processes vision) occupies the largest part of human brain compared to other senses. The size of an ostrich's eye is even bigger than its brain. Visual memories are the strongest memories. However, this privilege of sight over other senses can be misleading in concluding a fact. For an example, looking at an elephant's large ear, long tusks and small eyes, one can easily interpret that elephants are much better at hearing and chewing while they have a poor sense of vision. Actually, the elephant's ear flap is not for aiding in better hearing, rather it regulates body temperature, the tusks are meant for digging and holding, and the small eyes are good enough for sufficient vision. Many believe that eyes make mistake. But, I would say that "Eyes never lie; it is our brain, that interprets the image, makes mistake". The eye is just an image-forming device. It is our brain, which perceives the objects. Thus, mainly the brain controls visual perception. I will elaborate the image formation in the eyes in the subsequent chapter.

Evolutionarily eyes have been developed in animals mainly as a need to look for food. On the other hand, plants did not develop eyes as they did not need to move out or look for food far away as their basic nutrients are available close to them. One example that supports the view is that some cave dwelling animals like the salamanders (Olm) do not have eyes, as they do not need to see anything in the darkness of the cave. Depending on the food habit, different animals have developed different adaptations in their vision. Nocturnal animals can see in the very dim moonlight. To adapt for night vision they have bigger eyes, owls having the largest. Bigger eyes help in collecting more light to improve vision. Eagles have high density of photoreceptors (the pigments that receive light), which helps them to have better resolution of the sight from far. Animals like dogs, cats and foxes have a reflecting surface called "tapetum lucidum" behind the retina of their eye. Reflection of light from this surface back to the retina improves the quality of the image in the eye. Human eye can see only in the visible light whereas bees and some birds can see even in the UV light. Likewise, pit viper snakes can see in the infrared light. The bottom line is that animals need light to see. No light, no vision. Here I would like to say that light has helped develop vision in animals.

Why does the eye look black? By the way, why does anything look black? If an object absorbs all the colors of light, then it looks black. It means that a black paint absorbs all colors and reflects nothing. However, if a black object does not reflect any light, then how can we see it? Nothing is 100% black. A 100% black object cannot be seen. Blackness has also gradients. A black object reflects at least some percentage of light (e.g. 0.1%) that enables us to see. On the contrary, if an object does not absorb any light, it will look completely white. In other words, we see the color of an object from the light it reflects. A tomato looks red because it reflects red light. The central portion (pupil) of our eyeball looks black (figure 1A). When we describe the color of eye, it usually refers to the color of "iris"- the ring surrounding the pupil (figure 1B). People have different iris colors- black, blue, brown or gray etc. We can compare the eye as a blackbody, although not ideal. A perfect blackbody is an idealized hypothetical insulated cavity with an opening that allows only the entrance of electromagnetic radiation but does not allow the escape of any rays (Figure 1D). The eye's design is like that of a black body. It has a central transparent opening called pupil. The eye isn't black although it looks black. The inside of the eye is also very much transparent. An example can be seen in figure 1C where I show the picture of a Pompano fish's vitreous body that covers majority of the eye's cavity. The transparent cavity of the eye is protected inside the insulated wall of the eye so that light rays entering through the pupil do not escape from it (figure 1E). As light does not escape out from the eye, we cannot see the inside of the eye even if the pupil has a transparent opening because, to see something, we need light coming from that object. The eye doctor dilates the pupil and focuses bright light to see the inside of the eye. The typical cavity structure of the eye is ideal for image formation as the absorbed light inside the eye is used for visual perception. Nevertheless, we cannot see the image formed inside the eye. I will elaborate the mechanism of image formation as you read through the book.

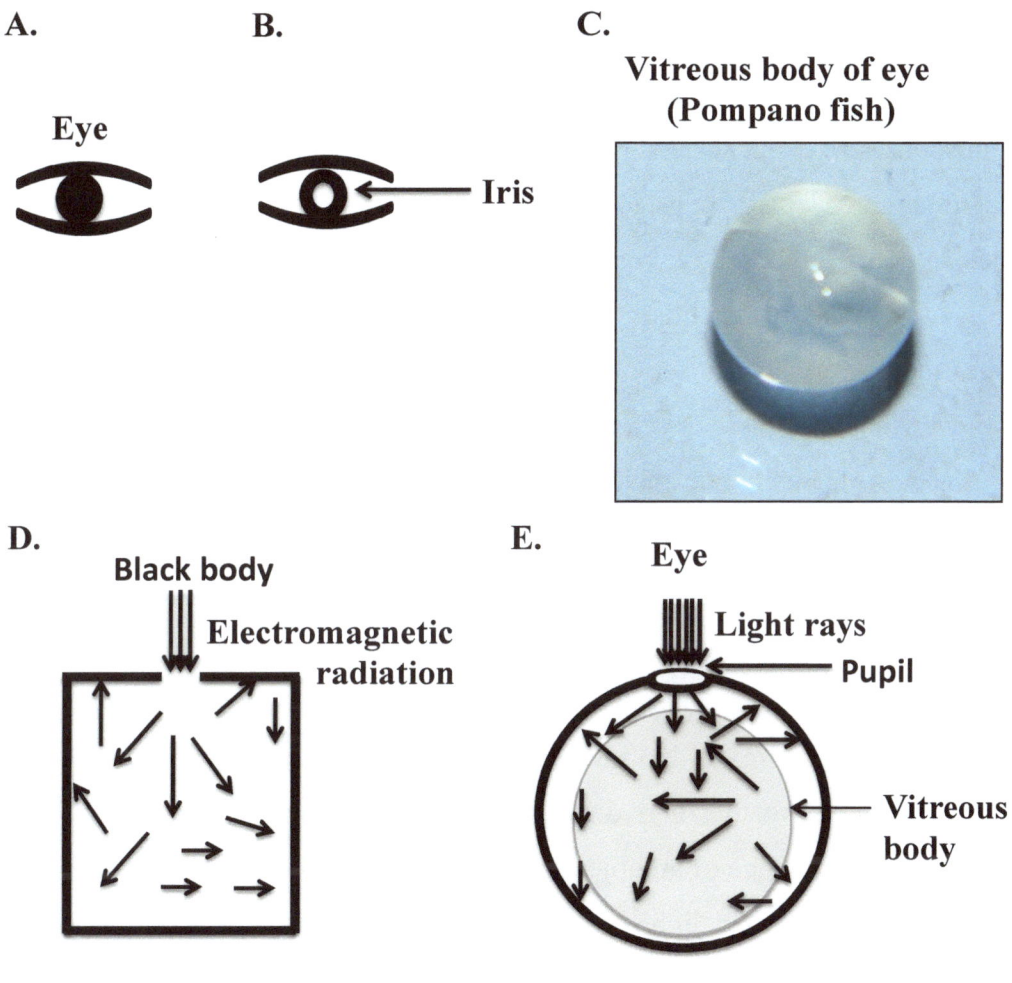

Figure-1

"Light is image and image is light".

Image

What would have happened if we had no eyes? Would there be an alternative way to see things? Before finding the answers let's ask another question. When we see things what do we actually see? There are two main theories to explain how we see- emission theory and intromission theory. Ancient philosophers like Euclid and Ptolemy believed in emission theory, which states that rays come out of eyes towards the object, which enables us to see. This ancient theory did not prove right over time.

The modern scientists have established the intromission theory, which states that we are able to see objects because light rays from the object enter our eye. Newton and others developed this theory in the 18th century. When we see an object we actually see the light coming from the object, it is our brain that perceives it as an object. Without our brain, we wouldn't be able to see. It's not that without eye we cannot see. Light and image are two sides of the same coin. We see the image because of light. Interestingly, an image is nothing but light. Wherever there is light, there is image and wherever there is an image, there is light. No light, no image. No image no light. We see an object because of light coming from it, whether it is direct or indirect (reflected or scattered). In other words, to see an image of an object, it must have light, either its own, like the sun (figure 2B) or reflected light from a source (figure 2A). For example, we see a tree in daylight because light from the sun falls on the tree which then after reflection enters our eye to form an image of the tree (figure 2). We can't see a tree in complete darkness, as there is no light. Even a shadow is light; it is just less light. Nobody can see without light. Even nocturnal animals can't see in complete darkness. They can see much better than diurnal animals but they do collect whatever dim light is available in the night to see objects in the dark.

A.

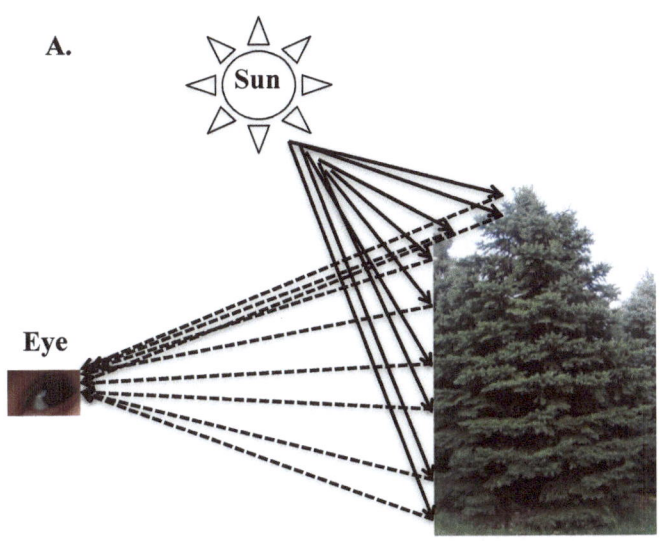

B.

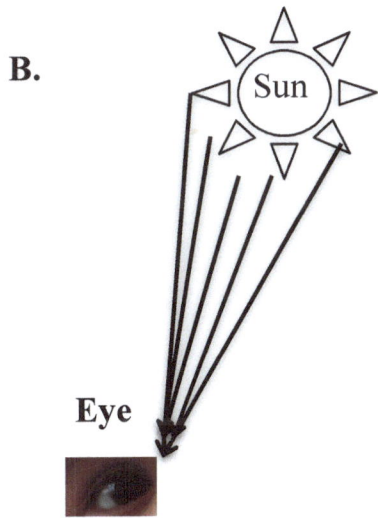

Figure 2

"Light helps us to see things but there is nothing that helps to see light".

Light

As we now know that image is nothing but light, let me describe how light helps in image formation. Before that, I would like to describe the basic properties of light. By "light" I would refer mainly to sunlight. Sunlight is the priceless gift of the sun to the earth. Human beings have been fascinated with sunlight through the ages. In many cultures, sun is worshiped as a God. It is not only the source of energy for the feasibility and sustainability of life but also an essential thing that gives vision to living beings on this planet. Vision is not only a basic need for searching food but also an important weapon of defense for organisms. Light helps in perceiving "vision" not only ascribed to the animals' sight through eye but also to the photo-sensing mechanism in blind animals like the cave dwelling salamanders, phototropism in plants and phototaxis in microbes. Although sunlight is the most visible thing that we see and that helps us to see things, there is nothing that helps us to see light. Light does not have any mass or physical identity.

Dispersion

Historically a lot has been discovered regarding the nature of light since the beginning of modern science (15th century). Earlier people thought that sunlight was white. Nobody had thought of light having many colors. Sir Isaac Newton used a prism and demonstrated that the white sunlight is made up of seven colors- red, orange, yellow, green, blue, indigo and violet. By the way, nobody knows why we see colors, light does not have intrinsic colors. A color is just the creation of the eye. Here is a simple illustration of colors in figure 3.

When I passed sunlight trough a glass prism the seven colors were dispersed. The colors were dispersed according to their wavelength, with the blue light (shortest wavelength) being dispersed the most, and red light (longest wavelength) being dispersed the least (figure 3A). This phenomenon of separation of individual colors from a polychromatic light is called "dispersion". To prove that the seven colors (ROYGBIV) combined to form white color back again, I passed the dispersed colored lights through a magnifying glass. A magnifying glass is a convex lens that converges light. When the seven colors of light merged back white light was formed (figure 3B). In this experiment, we saw that the white light is made up of many colors (polychromatic). We can imagine that the seven colors are not separate from each other, but they are mixed together as in figure 4. White light does not have a single wavelength but it is the mixture of wavelengths of all the seven colors. We don't know why our eye perceives the different wavelengths of light as different colors. Similarly, we also don't know why the seven colors of light merge together to form the white color. Again, it is our eye that perceives it as white color.

Rainbow

Yet in another way, we can observe that the sunlight is made of different colors is "Rainbow". Ancient people had a different perspective of the rainbow. In many faiths, the rainbow is considered sacred and divine. According to Hindu religion, the rainbow is the bona fide bow of the God of Rain "Indra". In ancient Japanese faith, it was believed as a bridge people used to descend to the planet. Even before Newton discovered that the white light is made of seven colors, people had known that a rainbow is formed when sunlight passes through raindrops. Usually, we see the rainbow in the sky during rain and it appears like a bow, as we cannot see the full view standing on the ground.

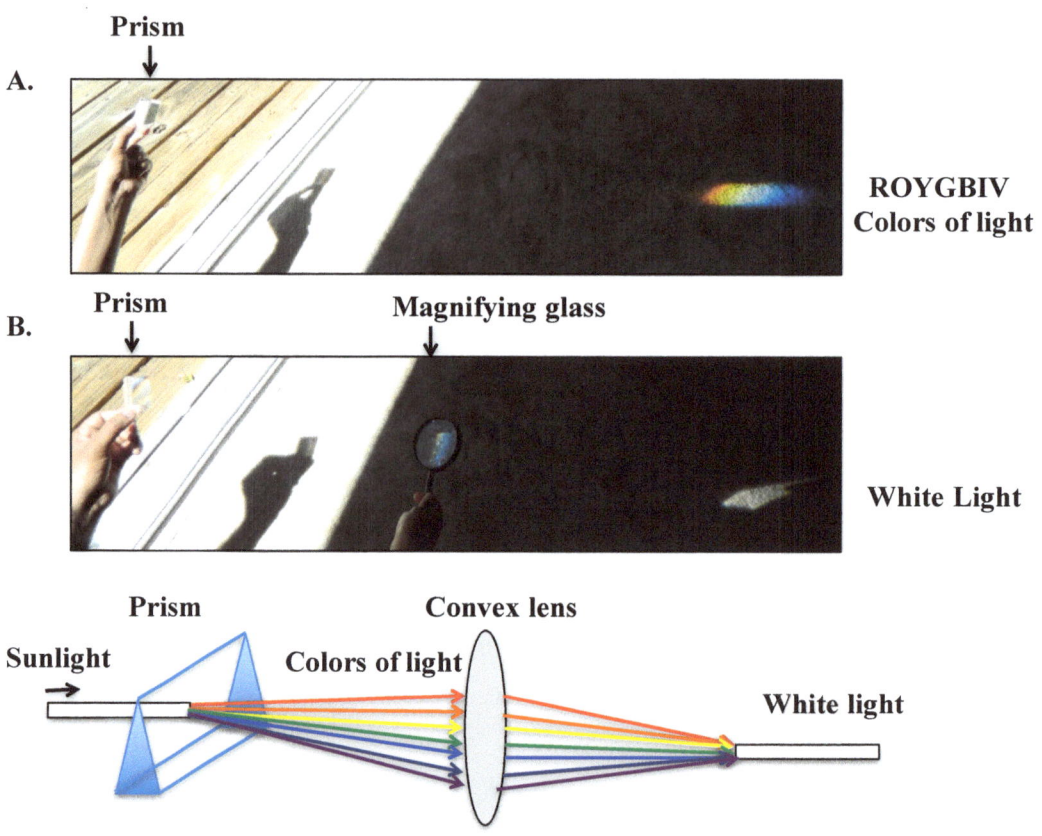

Figure 3

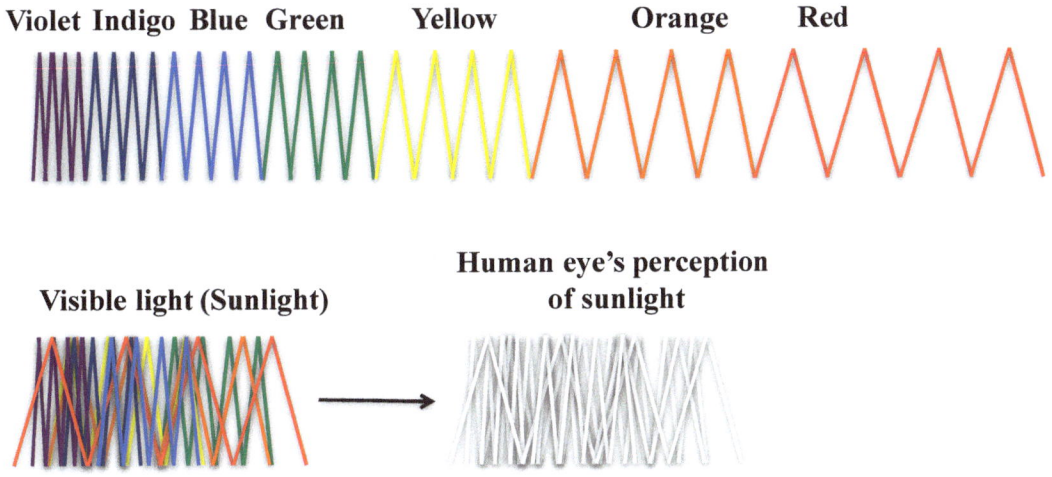

Figure 4

We can observe the full circle of the rainbow if we go above the ground. Once I observed the full circle rainbow when I viewed from the window of an airplane down the sky while it was raining during an afternoon. You can create a circular rainbow by spraying water from a garden hose in the afternoon (figure 5A). The phenomenon of the rainbow is a good example of the total internal reflection of light (I will describe it in more details later). When sunlight enters a raindrop, it is refracted twice- once when it enters the water drop and next when it exits the water droplet (figure 5B). We can think of a raindrop as a mini water lens. As an example of water lens, we often notice that while viewing through a transparent plastic bag full of water the image of newspaper letters appear enlarged. Water being a different refractive medium than air, we see the colors of light dispersed like a prism (figure 5B). When we look at the circle of all rain droplets that disperse colors, we can see the full circle of the rainbow. An example is shown in figure 5B where I show rainbow formation while spraying water from a garden hose during an afternoon. The different colors of light are dispersed differently after coming out of water drops- red being dispersed the least, it appears on the top of the rainbow and blue being dispersed the most, it appears on the bottom of the rainbow. Thus, different colors being dispersed at different angles, the cone of light from the observer's eye to the circular rainbow has ~42 ° semi angle. We can see the full circle of the rainbow by taking pictures of the different sides of circles as in one full view the full circle was difficult to capture in a still image of the camera but apparent in the panoramic movie. By the way, the rainbow is a virtual phenomenon, we can see it in the eye, but cannot project into a screen, as the rainbow is not a real object. It does not appear at a fixed place rather it is a virtual object (an optical illusion) that appears from our eye's perception of the dispersed colors.

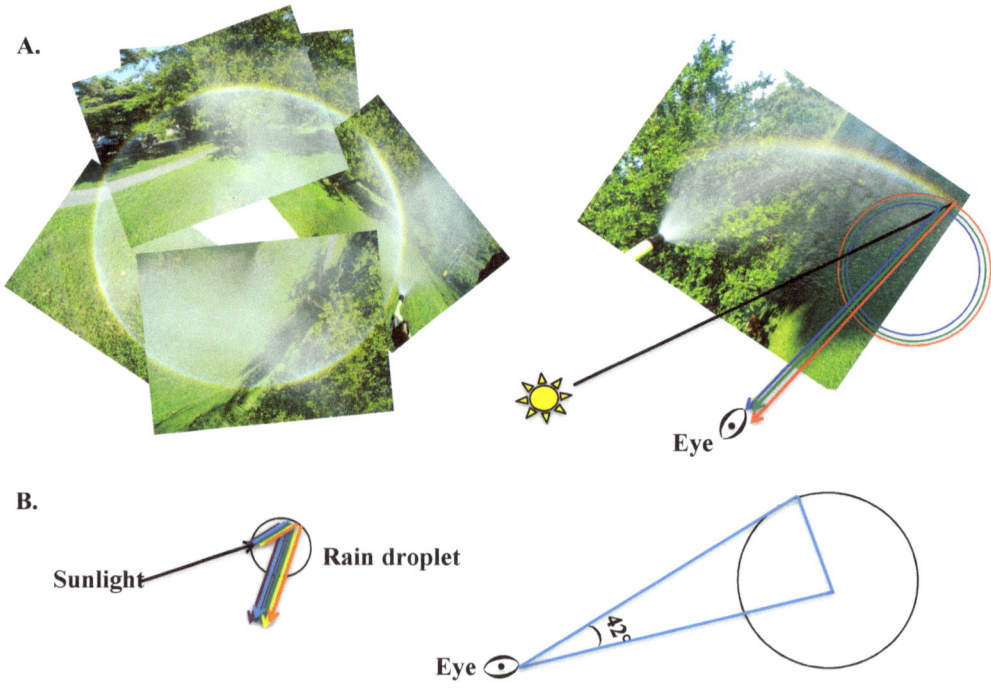

Figure 5

"We are faced with a new kind of difficulty. We have two contradictory pictures of reality; separately neither of them (wave or particle) fully explains the phenomena of light. This interpretation (wave particle duality) ...appears to me as only a temporary way out".

- *Albert Einstein*

Wave-particle Duality

The most intriguing question pertaining to light is whether it is a particle or a wave? For centuries, scientists have been struggling to understand this paradox- what exactly is light? Surprisingly in some contexts light behaves as particle while in others it behaves as wave. Let me first explain the wave nature of light, as it is the classical view. The particle nature of light is accepted as the modern view.

Light as a wave

Christiaan Huygens proposed the wave theory of light in 1678, which was further established by scientists like Descartes, Robert Hooke, De Broglie, Thomas Young and Maxwell. In the absence of a real picture of what the light looks like, we can think of light moving like a water wave. Usually, we imagine the picture of the sun as radiating light rays in the straight line as depicted in figure 6A. If we consider light as a wave, we can think of the sun as a point source producing light in spherical wave front outward in all directions in the solar system (figure 6B). Unlike water wave or sound wave, light does not need any medium to propagate. In other words, it is self-propagating. As it moves

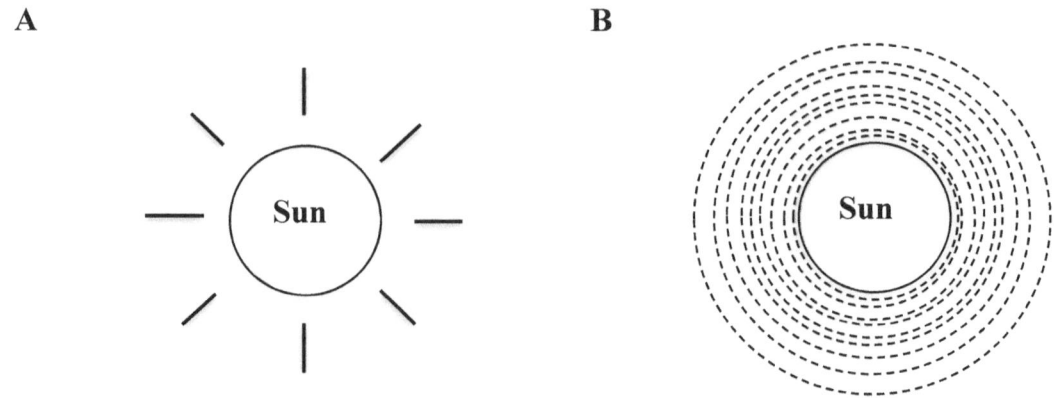

Figure 6

in the space or vacuum, it moves in the medium like air, water or glass too. Like X-ray, UV ray, microwave and radio wave, the visible light is an electromagnetic radiation (EMR)- it propagates with both electric and magnetic field oscillating perpendicular to the direction of propagation. By visible light, I want to say the portion of sunlight that we reach the earth's surface after atmospheric absorption and that human eye can see. The sun, like a black body, radiates entire spectra of electromagnetic radiation. Most of the non-visible portions of light are absorbed by the earth's atmosphere. In 1865 Scottish scientist James Clark Maxwell proposed that light is a form of electromagnetic radiation, which was later confirmed experimentally by Hertz. Maxwell in his work- "A Dynamic Theory of Electromagnetic Field" published his famous equation of electromagnetic wave. His work demonstrated that electromagnetic radiation could move at the speed of light in space.

Electromagnetic radiation

Any radiation (including visible light) that has both electric and magnetic field associated with it is called electromagnetic radiation. Electromagnetic radiations can have the wavelength from as small as one hundred millionths of a meter (gamma rays) to one hundred kilometers (radio waves).

Electromagnetic radiation can be divided into three basic type of radiations – the shorter wavelength (<380 nm) region includes U.V. rays, x-rays and gamma rays which are expressed as their energy level, the visible light radiations (between 380 to 760 nm) which are expressed in terms of their wavelength and the higher wavelength radiations (>760 nm to m or km) like the microwave and radio wave which are expressed in terms of the frequency. In reality, there are no strictly defined boundaries between different radiations. A wavelength is a distance between two subsequent crests or troughs of a wave (figure 7). So, one wavelength is one complete cycle. Frequency is the number of cycles per second, which is expressed as Hertz. When we listen to the radio with AM or FM we notice this number. The radio wave frequency for AM (amplitude modulation) is expressed in kHz and that for FM (frequency modulation) is expressed in MHz. For an example, radio- 98.3 FM indicates the frequency of 98.3 MHz. In simple words, if you stand at a point and observe the radio wave (we cannot see it though), it will pass 98.3 million complete cycles (or wavelengths) per second. I will define the cycles of the light wave later in this section.

Frequency and wavelength are related to the speed of light. Frequency is the speed of light divided by its wavelength. For example, to calculate the frequency of red light of 600 nm, we can divide the speed of light in vacuum ~300 million meters per second by the wavelength- 600 nm, which comes out as 500 trillion cycles per second. Thus, the frequency of red light is ~500 terahertz. This is way bigger in number compared to the frequency of radio wave. Light with shorter wavelengths has the higher frequency and vice versa. Likewise, light with higher frequency has the higher energy. Thus, blue light has more energy than red light. On the other hand, longer wavelength light (red light) can penetrate deeper into the medium.

Different electromagnetic waves vary in their frequency, wavelength, and energy, but they do not vary in one thing- their speed, which is the "speed of light". The speed of light is constant- exactly 299, 792, 458 meters per second (approximately 300 000 000 m/s). For example, a visible light of 600 nm wavelength multiplied with its frequency 500 trillion Hz will result in the speed of 300 000 000 m/s. On the other hand, a radio wave of 300-meter wavelength multiplied with its frequency of 1MHz will result in the same speed as red light of 300 000 000 m/s. The sunlight takes ~8 min to reach the earth from the sun. So, when we see the first light rays during sunrise, we are actually 8 min behind as the sunrise has happened 8 minutes earlier. The same time (8 min) will be needed whether it is visible light or radio wave to reach the earth from the sun. The sun's light has all the spectrum of EMR- radio wave, microwave, visible light, UV, X-ray etc. and they all take the same time to reach the earth's surface where all except the visible light are absorbed by the outer layer of the earth's atmosphere.

Let's compare the radiation of different wavelengths and frequencies to some simple examples. A tractor with big wheels at the speed of 30 miles per hour will reach the same distance at the same time as a car with small wheels can go with the same speed of 30 mph. However, the tractor's wheel will go less number of cycles or round per minute (rpm) as it is big compared that of the car. If we compare the diameter of the wheel (width) as the wavelength (or a cycle) and the rounds per minute as the frequency, then both the tractor and car will have same speed despite having different wavelength and frequency. We can also compare this to a race between a giraffe and a rabbit both running at same speed, let's say 20 miles per hour. The giraffe having longer legs will step much fewer times than the rabbit to reach the same distance. If we imagine the number of steps as the frequency and the distance between two subsequent steps as the wavelength, then the giraffe and the rabbit will have different frequencies and wavelength but the same speed and will take the same amount of time to reach anywhere.

Sine Wave

Although sunlight's elementary particle "photon" has no charge or mass, it shows electrical and magnetic property upon interaction with charged and magnetic body respectively. Light propagating with no net charge can be well explained if we view it as a wave having an oscillation of the electric and the magnetic fields perpendicular to the direction of propagation. A moving electric field produces magnetic field and a moving magnetic field produces electric field. Light is usually thought of as a sine wave as demonstrated in figure 7B. A sine wave is a periodically oscillating wave in which particles oscillate in a smooth repetitive manner like a simple pendulum. The distance between two consecutive crests or troughs is called wavelength. The brightness of light wave is measured in terms of its amplitude, which is the height between resting state (0, X-axis) to the highest point of oscillation- crest or trough (+1 or -1). Let's assume a particle (red dot) moving from position zero towards one and back to its origin in a unit circle as described in figure 7A. The particle's displacement can be expressed as a function of sine theta, which are basically 0, 1, 0 and -1 as shown in the unit circle. When the particle completes one whole cycle it comes back to zero (whether it goes from 0 to 0 or 1 to -1 and then to 1). Its net displacement or total sine value becomes zero. In the same way, oscillating electric or magnetic field can have no net electric charge or magnetism if light follows a sine wave pattern.

Sine Wave

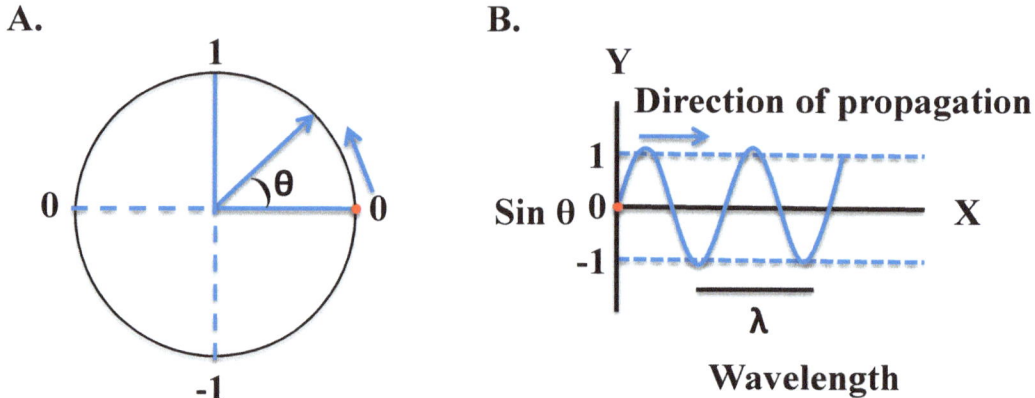

Figure 7

Refraction

Light changes direction when it moves from one medium to another. This process is called "refraction" (figure 8). The degree of bending of light depends upon the refractive index of the medium. Refractive index (n) is defined as the speed of light in vacuum divided by the speed of light in the medium. When light moves from a high refractive index (glass) to low refractive index medium (air), it bends away from the normal (figure 8, left panel). On the other hand, when light moves from low to high refractive index medium it bends towards the normal (figure 8 right panel). This change in direction can be well explained if we think light as a wave.

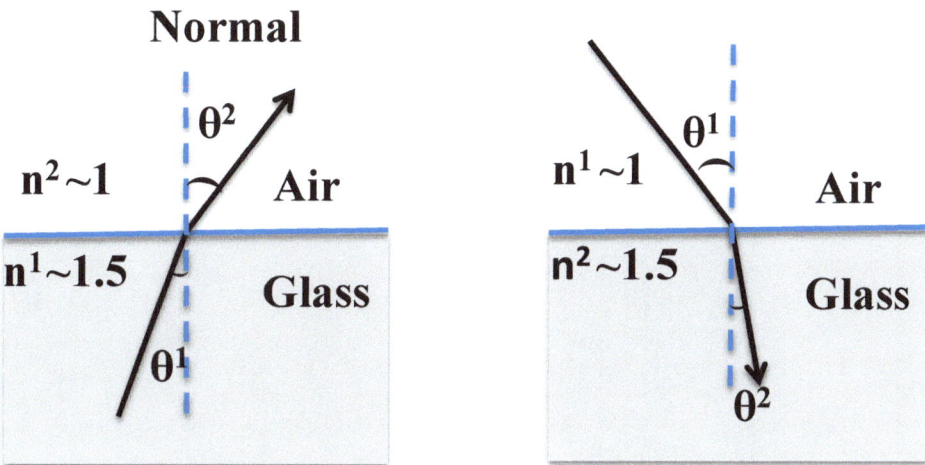

Figure 8

Diffraction

The most important behavior of light that proves its wave nature is "diffraction". The phenomenon of spreading out of light when it passes through a small aperture is called diffraction. Thomas Young in 1801 performed the double slit experiment and showed that when light passes through two parallel slits, alternate bands of bright and dark pattern appears on the screen behind the slits. This can only be explained if light can be thought of propagating as parallel waves that while passing through the slits spreads out as circular waves, which interfere with the wave from another slit (figure 9). If light were particles, they would go straight through the slits and should not produce interference like waves. Diffraction is very important to understand the concept of microscopy and the limit of resolution. I will explain it in more detail in the later part of this book.

Double Slit Interference

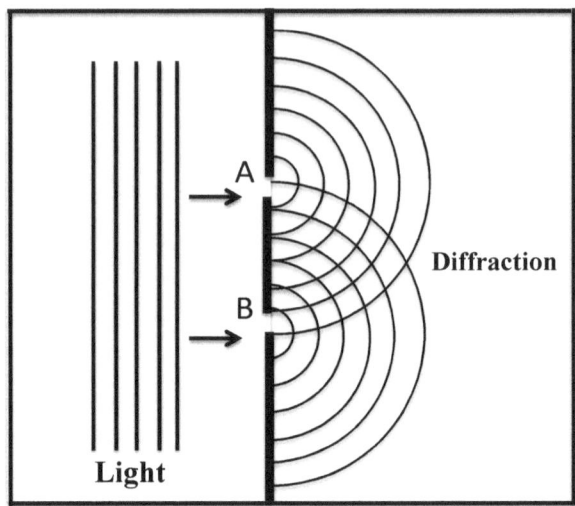

Figure 9

Interference

      Interference is a phenomenon that occurs when two waves meet each other (figure 10). As shown in figure 10, we can think of interference of two waves– blue and red when they interact. In the left panel, the two waves are in phase whereas in the right panel the two waves are out of phase. Two waves are said to be in phase when the crest of one wave aligns with the crest of another. In other words, oscillating particles of both the waves are exactly at the same position of the unit cycle at any particular moment of time. In this case, when two waves in phase meet each other (figure 10A and B), the net amplitude of waves increases and thus it results in bigger wave (in the case of water wave) or brighter wave with higher amplitude (in the case of light) (figure 10C). This is called constructive interference. This is exactly when crest (+1 in unit circle) of one sine wave adds to crest (+1 in the unit circle) of another sine wave, the net amplitude becomes 1+1=2. Contrary to this, when two waves are out of phase, the crests of the two waves do not align, rather crest of one wave aligns with the trough of the second one. In other words, oscillating particles of both the waves are not exactly at the same position of the unit cycle at any particular moment of time. In this case, when two waves out of phase meet each other, the net amplitude of waves becomes zero as the crest of one wave cancels out the trough of 2nd wave (figure 10D and E) and thus it results in no wave (in case of water wave) or darker bands (in case of light) (figure 10F). This is called destructive interference. This is exactly when crest (+1 in unit circle) of one sine wave adds to trough (-1 in the unit circle) of another sine wave; the net amplitude becomes 1-1=0. Of course, there will always be different extents of interference depending upon the level of the phase difference.

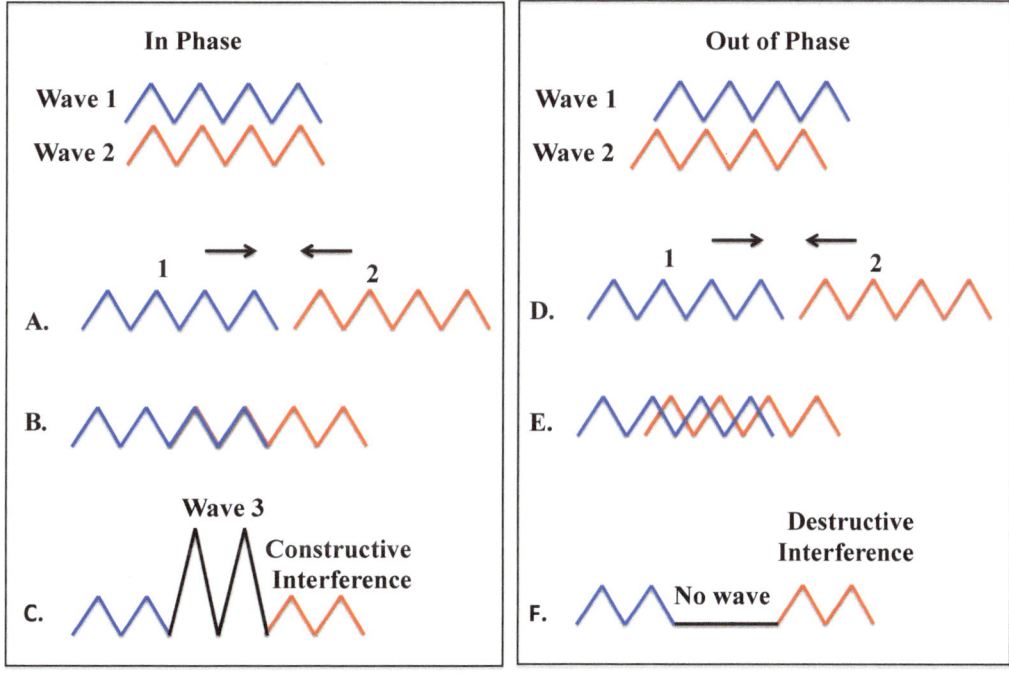

Figure 10

We can create double slit interference using simple tools like a laser light, cardboard tube and black construction paper (figure 11). We can make two parallel slits on one side of the cardboard tube using black construction paper and a thick copper wire as described in the figure 11 A. Next, if we focus the laser light through the slits we can see the interference pattern on the wall (figure 11B). We see the alternate bright and dark bands of laser light because of the interference of the light waves coming out through each slit. The brighter bands result from constructive interference and the darker bands result from destructive interference. Dark band means no light. It is the region of no wave of light. Instead of a laser, we can use any parallel light, but the light from ordinary electric bulbs does not produce parallel waves because of the smaller circular front. When sunlight reaches us it nearly becomes parallel because the sun is far away from the earth. When we look at sunlight from the earth; we actually look at a very tiny portion of sun's circular wave front, which appears almost parallel.

We can perform similar two-slit interference experiment with sunlight using a cardboard box and black construction paper. We can make two parallel slits as shown in figure 12A. Next, using the side hole if we look at the sunlight coming through the parallel slits it will appear as colored bands alternating with dark bands (figure 12B). Because sunlight is made up of seven colors and each color having different wavelengths, they diffract in different angles. In the earlier example of interference with the laser, the laser light was a single color red wavelength light. Therefore, we did not see color bands in the interference pattern. However, sunlight is polychromatic, thus we see the colored bands in the diffraction pattern. Now we know that diffraction of light is related to its wavelength. Light with short wavelength will have less diffraction. The dark bands are the regions of no light. The brighter bands are the region of constructive interference.

Another important parameter that proves light to be a wave is polarization. Polarization is basically the direction of oscillation of wave. When we look at the waves in an ocean, we see that water wave oscillates vertically to the direction of propagation. Sound waves oscillate longitudinally (along the direction of propagation). However, light can have multiple polarization- plane or circular or elliptical. Wave theory of light is more suitable to explain the polarization behavior of light.

We can assume light as an electromagnetic wave, but there are certain behaviors of light that does not reflect its wave nature, which I will discuss in the particle theory. Nevertheless, light interacts with molecules much smaller than its wavelength.

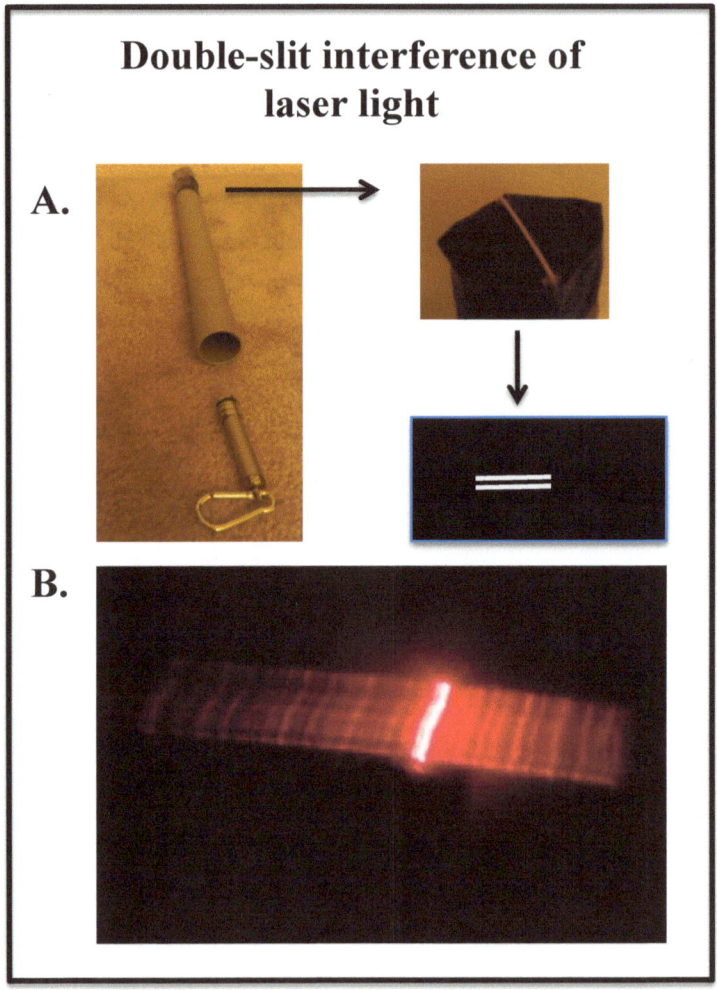

Figure 11

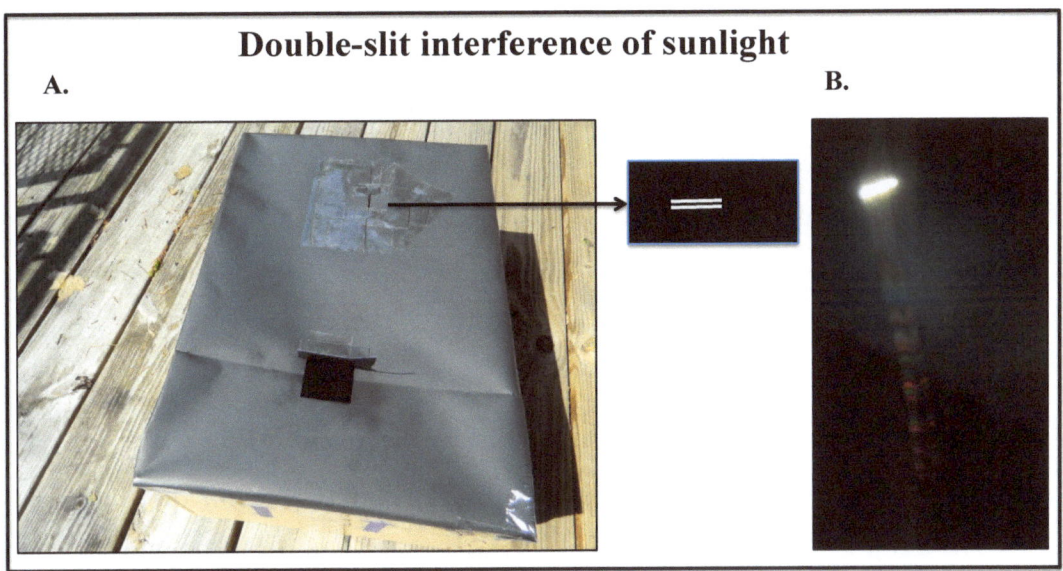

Figure 12

Light as a particle:

Now let's look at the particle view of the light. The particle theory was developed by Newton, which was further established by scientists like Bohr, Planck, and Einstein. This theory is more accepted, as it is the modern view that explains the behavior of light that cannot be explained by wave nature.

Reflection

Newton mainly developed the corpuscle theory by his observation that light reflects in a straight line (figure 13). Bending of light back into the same medium from which it came after hitting a reflecting surface is called reflection. Newton believed that if light were not made of corpuscles (mini-particles), it would not reflect in straight lines. According to the laws of reflection, the angle at which incident rays arrive at the normal drawn at the surface is equal to the angle reflected rays go away from the normal (figure 13).

We can observe the phenomenon of reflection in a mirror, which has a perfect reflecting surface. When we look at the image in a mirror, we actually look at the projected image (virtual) of rays reflected from our body. On a side note, interestingly when we look at our image in the mirror, our left-hand looks right and the vice versa. We may assume that the image in a mirror appears left-to-right inverted. Why does the left-hand look right in the mirror? It is actually not inverted. The mirror image is a direct reflection of lights (front to back inversion) from the object. That's why it just switches front to back. It is our brain that interprets as left and right considering the axis of symmetry of image. It is as simple as when we see another person in front of us. His left eye is at our right side and vice-versa. We don't notice this difference when we place two objects in front of a mirror.

Reflection

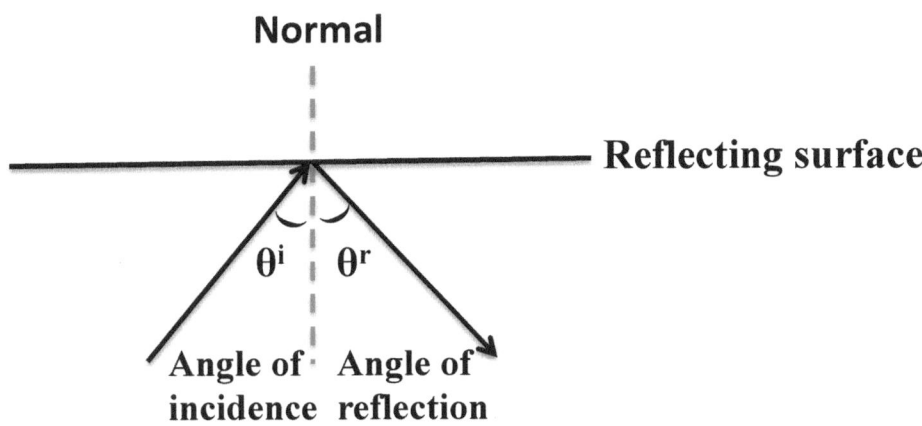

Figure 13

Photoelectric effect:

The particle theory subsided for a long time (since Newton) as the wave theory by Huygens prevailed till the Nineteenth century when Scientists like Max Planck and Albert Einstein developed the particle theory of light. Planck proposed that the radiation of blackbody is not continuous (like a wave), rather quantized and made of discrete units of specific amount of energy. This quantized unit of energy is called "quantum". Quantum (in Latin it means how much) is the smallest unit of the physical entity that is involved in an interaction. Einstein extended Planck's model to sunlight and proposed that the sunlight is also quantized in the form of discrete units of energy. The discrete units of energy of light are called "photons". Simply, a photon is the quantum of light. The photons have a specific amount of energy, $\{E\}=h.v$, where v is the frequency of light and h is Planck's constant. German scientist Heinrich R Hertz (the unit of frequency "hertz" is named after him) observed in 1887 that metal electrodes exposed to UV light could produce an electric spark. However, Hertz did not think of the mechanism behind his observation. A technician in the patent office (hitherto infamous Albert Einstein) explained this phenomenon in his proposed theory of photoelectric effect in 1905 for which he was awarded the Nobel Prize in 1921. Einstein explained that light behaves as particles and has discrete units of energy called the photon. A photon can knock out an electron from any material (more observable in metals) however, only a photon of specific energy level (photon energy, PE=hν) can knock out specific electrons from metals (figure 14, left). As in Hertz's experiment, the UV light's photon but not the red light's photon had required energy to knock out electrons from the particular metal used.

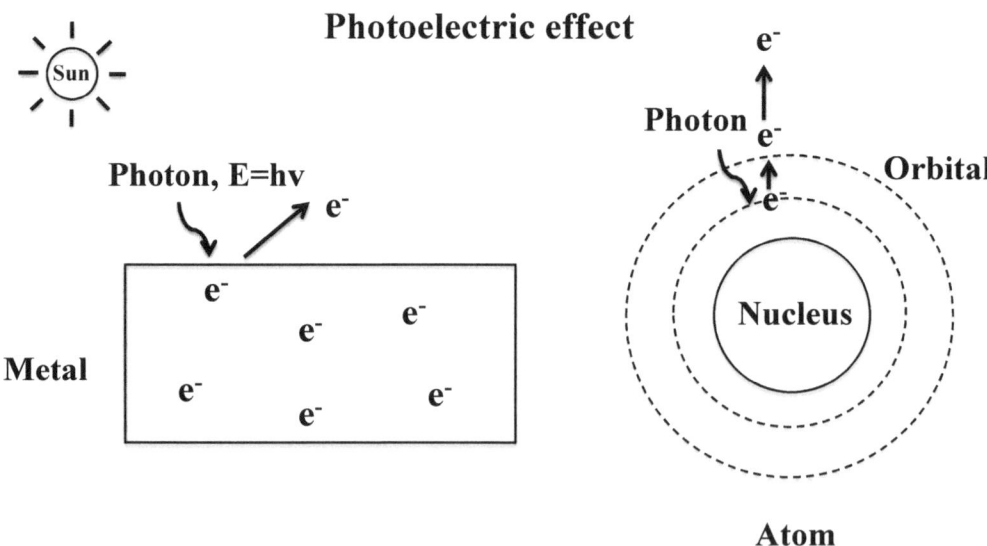

Figure 14

The energy of a photon is dependent on the frequency and wavelength of light but independent of the intensity. The high intensity of low energy of photons is not equal to the low intensity of high-energy photons. If that were the case, then the high intensity of long wavelength red light would be equivalent to the low intensity of blue light. For an example, let's imagine that an electron from a metal needs 5eV energy to be knocked out of the orbital. If we assume that a single photon of blue light has 5eV. In this case, 100 photons of blue light can knock out 100 electrons from the metal. On the other hand, let's assume that a photon of red light has 2.5eV. If energy were continuous (like a wave) then, 200 photons of red light would sum up to 500eV (200X2.5eV). In this case, twice the amount of red light (than the blue light) should be able to knock out same 100 electrons. However, it is not the case. Actually, the red light will knock out zero electrons as the photon energy is below the threshold energy level (5eV). A spinning electron in an atom is associated with a certain level of energy in the orbital (figure 13B). A photon below the threshold energy level cannot knock out the electron.

However, there is an exception to the above-mentioned concept. An atom that is excited by blue light photon can be excited by two red light photons of half the energy if both photons of same phase excite the same atom simultaneously. This has been possible after the invention of LASER light that produces high flux of stimulated emission photons. This concept is utilized in two-photon microscopy to image deep tissue using higher wavelength light.

The fact that transfer of energy is also not continuous and occurs in quantized form cannot be satisfied with the wave nature of light. Prior to the Einstein, it was believed that light is a wave and needs an aether like medium in space to propagate. Einstein proposed that the concept of aether as a medium of propagation of light in space is not right. Rather he stressed that there is no such medium in space. Light does not need any medium to propagate in space. He also proposed in the special theory of relativity that the speed of light is same for all observers in the vacuum no matter what the speed of source is. This was later proved experimentally. Einstein also proposed in his general theory of relativity that light, despite having no mass, can be affected by gravity. This was also proved later by several mechanisms including gravitational lensing which shows the image of stars (like twin QSO) upon bending of light by the gravity of galaxies.

Quantum revolution was brought forward by scientists like Planck and Einstein whose particle theory contradicted the wave theory of light and dominated by it. However, it caused a new dilemma in understanding the nature of light- wave or particle? Surprisingly, many scientists, later on, reported wave like behaviors of particles. One notable example is the experiment by English scientist GP Thomson who published his observation in 1927 that electron (which is a particle having definite mass) can also exhibit wave like diffraction pattern. Using a beam of an electron (cathode rays) passing through a thin metal film and capturing the image on photographic film screen a centimeter away he observed the diffraction rings around the central beam. Other particles like proton and neutron or even bigger particles were later on shown to be exhibit diffraction pattern like a wave. Other experiments with interferometer revealed dual nature (both wave and particle at the same time) of light. All of these observations led to the development of "Wave-Particle Duality Theory" which is more and more being accepted recently as it can only explain the dual nature of light. We can think of all the particles moving as a wave. For example, the orbital path of electrons around the nucleus of an atom can be thought of as waves. It (nature of light) still remains an open question for future research.

"Light is image and image is light".

Now that we have discussed some of the basic properties of light, we can perceive how light helps in image formation. All of the physical properties of light are relevant concepts for microscopy as they all contribute to different aspects of image formation from human eye to pinhole camera to superresolution microscope. No matter how we view light as some interactions of light with matter are well explained by its wave nature and some are well explained by its particle nature. I will explain how these basic properties of light are helpful; how refraction helps in image formation, how wave nature of light helps or interferes in viewing smaller objects and how particle nature of light helps in processing the image.

The image is nothing but light coming from the object. When we look at sunlight passing through the roof, we actually look at the image of sun. Light is essential for objects to be visible. We cannot see blackholes because light does not escape from them. Black is not a color. It does not have any wavelength. It is a just absence of light. A black paint does not produce or reflect any light. A red tomato looks red because when exposed to light, red wavelength from tomato comes to our eye. Anything that produces or reflects light can be seen (explained earlier in this book). We just need an aperture to form image. Be it our eye, or hole in the window or roof, or pinhole or objective lens of microscope, they all have apertures which help in image formation. The resolution of an image is strictly related to the aperture size.

Pinhole camera

Sometimes I wonder if vision would be possible in the absence of the eye. I think the answer is yes. Then, what would be the alternative way to see without the eye? Let me explain this after I explain the concept of pinhole camera. If we think of the eye as a camera that captures the image, it is similar to a pinhole camera.

A pinhole camera is not a typical camera we see. It can be as simple as an insulated box with a small hole in the front to allow light to enter (figure 15A). It does not have any lens. We actually don't need a lens to form an image. I will explain in the later section how lens improves the image quality. If we place an object in front of the pinhole, an image can be formed inside the camera (figure 15B). The light rays coming from the top of the object will enter through the pinhole diagonally to form the bottom part of the image and the light rays coming from the bottom of the object will enter through the pinhole diagonally to form the top part of the image. In this way the image formed inside the pinhole camera will be inverted but real image. A real image is an image that can be projected onto a screen. If we place a photographic film or chips on the backside of pinhole camera, we can collect the image. This is how some pinhole cameras are designed without lens. Usually the resolution of pinhole cameras are poor. However, the aperture size and exposure time can significantly improve the image quality in pinhole camera.

A pinhole camera is not a perfect camera and cannot be used to generate very high-resolution image. But, the sharpness of the image depends upon the size of the pinhole and the distance of object from the pinhole. However, there is a limit to how small the size of pinhole can be, below which the details of a picture is lost. By the way, light collected by the pinhole camera is very small.

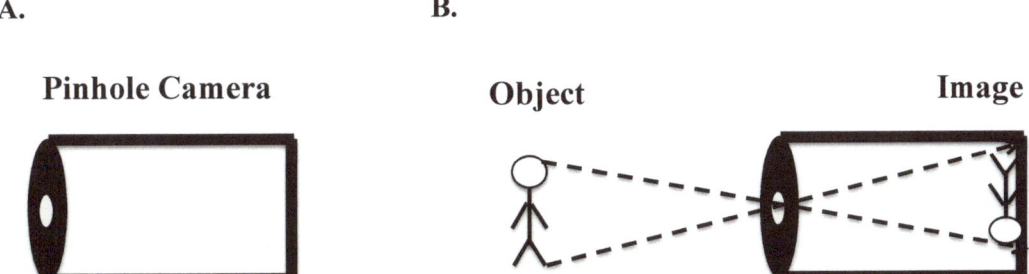

Figure 15

Thus, the resolution of the image is not very satisfactory. An example can be seen in figure 16. I made the pinhole camera (as shown below, figure 16 left panel) by taking an empty Pringles potato crisps container and opening a small hole on the back side. When I placed this pinhole camera in front of a light bulb, I could see the image of light bulb inside the Pringles container- a pinhole camera (figure 16, right panel).

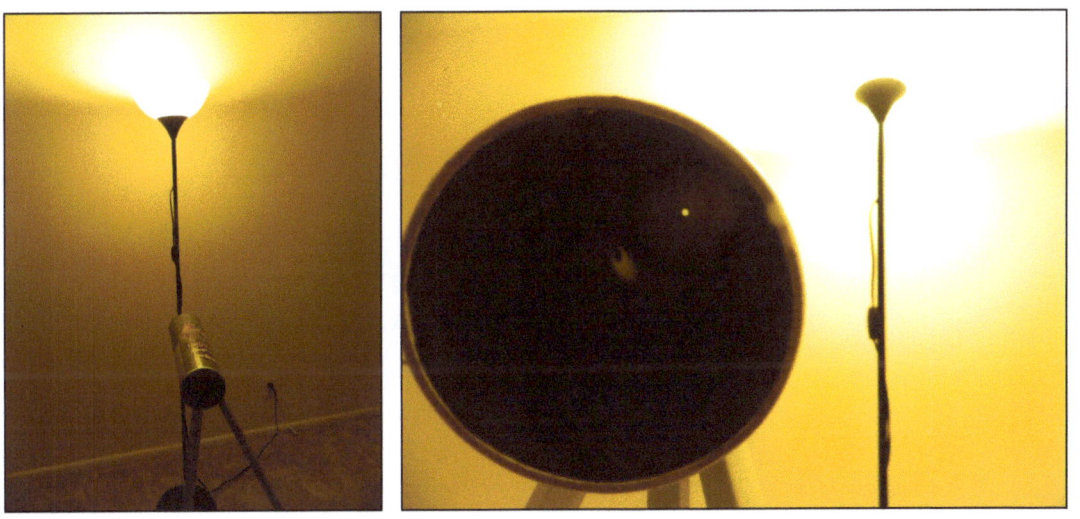

Figure 16

The human eye can be correlated to a pinhole camera. The pupil of the eye is like a pinhole, through which light enters the eye (figure 17A). Light from the object after passing through the pupil forms a real inverted image on the back side screen of the eye called "Retina" (figure 17B). The diameter of the pupil regulates the amount of light that enters the eye. That's the reason our pupil becomes smaller when exposed to bright light and bigger when exposed to dim light. The pupil of the owl's eye is much bigger to collect the maximum of dim light. But, unlike a pinhole camera, our eyes have a lens which helps in focusing light to form a sharp image on the fovea of the retina.

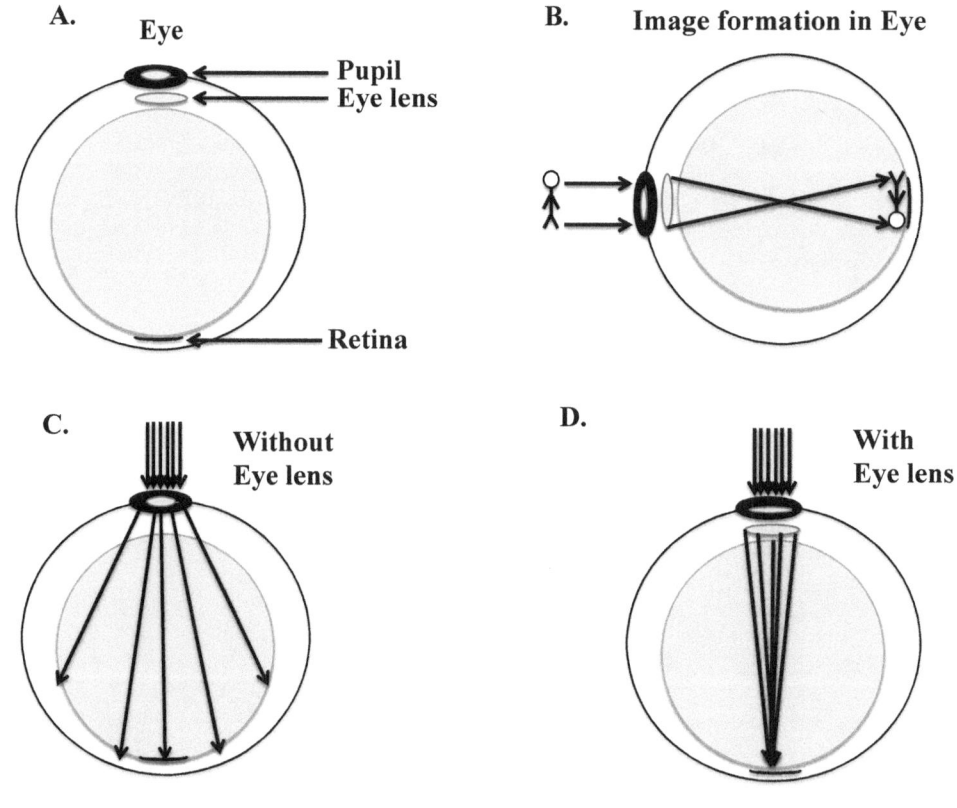

Figure 17

Let's imagine what would happen if there were no lens in the eye. I assume that an image can still be formed on the retina, but it will be very blurred as the light will not be focussed at a point (figure 17C). Because of the presence of eye lens, the light coming from the object is focussed to the fovea; the region of the retina where the light receiving cells (cones and rods) are concentrated (figure 17D). Now we can think that the eye without the lens would be like a pinhole camera. Therefore I predict that if humans or animals had no typical eyes, in other words, if evolution had not designed eye like structure, but designed a hole (like a pinhole) on the back side of which light receiving pigments were present, then vision would still be possible. This alternative mechanism of vision wouldn't be as efficient as the one with a typical eye. We would just be able to see or sense different kinds of objects. We can imagine such hypothetical eye to look in part like "Pit cavity" in the viper snakes that help in infra red sensing. Although a pit cavity is not an eye like structure, the pit membrane with receptor helps the viper's eye sense the infrared light that human eye cannot see. It (hypothetical eye) can also be thought of a simple hole like an ear with retina inside or like rudimentary eyes inside the skin of salamanders.

One of the main differences between a pinhole camera and eye or conventional cameras is the presence of the lens. So, let me explain in the next chapter how a lens improves the quality of the mage.

Lens

The word "lens" is derived from Latin name of lentil. The scientific name of lentil plant is *"Lens culinaris"*. Both lens and lentils have one thing in common, the shape- convex on both sides. Lentil's shape is convex like a lens. Nonetheless, lenses are available in many different shapes- convex, concave, plane, spherical etc. Here, I will describe mainly about the convex lens whose both the surfaces bulge in the middle. The eye's natural lens is convex type lens although it is not made of glass, it has a higher refractive index (~1.4) compared to that of water or other cells (~1.3). Lenses have been used by human beings since ages. Significant importance of lenses was felt since seventeenth century onwards after the "Father of Science" Galileo developed the telescope to observe remote stars and Robert Hooke developed microscope to observe cells. Be it a telescope or microscope, lens basically does the same function. Let's discuss why we need a lens to see better.

Refractive Index

The very important nature of lens is its refractive index. Refractive index (RI) is the numerical definition of reduction of the speed of light in the medium. RI can be calculated from the ratio of speed of light in vacuum to the speed of light in the medium. ($\mu=c/v$, where μ is the refractive index, c is the speed of light in vacuum and v is the speed of light in the medium. A glass lens usually has a higher refractive index (~1.5) compared to air (~1) and water (~1.3). A glass of RI 1.5 means that light moves 1.5 times slower in that glass compared to the the speed of light in vacuum. RI not only tells us about the speed of light but also determines the degree of bending of light in the medium. We have discussed it in the previous chapter that when light moves from low RI medium (air) to high RI medium (glass), it bends towards the normal (figure 8 and 18). This degree of bending follows Snell's Law which is as written as- $n^1 \cdot \sin \theta^1 = n^2 \cdot \sin \theta^2$, where n is the RI of the medium and θ is the angle of light to the normal. When light moves from a medium of RI, n^1, to the second medium of RI, n^2, then the angle at which light bends in the second medium (θ^2) can be calculated by the formula $\sin \theta^2 = n^1/n^2 \cdot \sin \theta^1$. We can apply this formula to determine the degree of bending of light after passing through the lens.

Why does a convex lens converge light?

Despite having same RI, light shows opposite angle of deviation in the convex vs concave lens. Whereas a convex lens converges light rays, a concave lens diverges light rays. Because of converging nature, the convex lens is used in the microscope. Let's analyze why light converges after passing through convex lens (figure 18). Two parallel light rays 1 and 2 after passing through a convex lens, will merge at the focal point of the lens (figure 18A). Ray 1 bends twice; once when it enters the glass (figure 18B) and then when it exits the lens (figure 18C). If we draw a tangent line at the point where light enters the glass, we can draw the normal axis which is the line perpendicular to the tangent line (figure 18B). The angle of incidence and refraction (deviation) are the angles between light and the normal axis (θ). A convex lens has convex surface because of which the tangent bends towards the lens, where as in a concave lens, it behaves the opposite. We can easily calculate the degree of bending by Snell's law as mentioned in the figure. Overall, if we calculate the angles θ^1, θ^2, θ^3, θ^4 light will finally bend towards the normal.

The second ray (2) that passes through the center which is basically along the normal, will pass straight (figure 18D), although its speed is as slow as the first one. Why does the light not bend when it passes along the normal? The answer lies again in the Snell's law. This ray enters the lens at an angle perpendicular to the tangent line, that means it is 0 degree to the normal axis, thus the value of sin 0 becomes equal to zero (figure 18D). This means that the angle of refraction also becomes 0 degree which results in sine value of zero. Sine value of zero means no deviation. For this reason, the light entering through the center of the lens does not deviate.

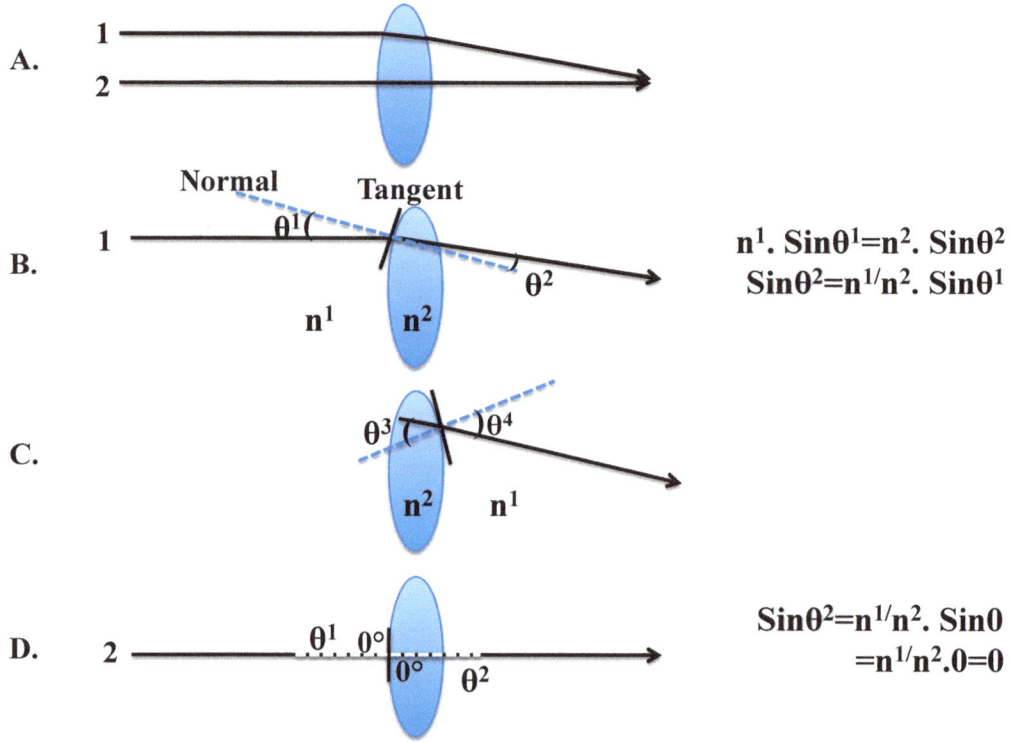

Figure 18

Now we know that a convex lens converges light after the later passes through it (figure 19). The point where all the parallel lights converge after passing through the convex lens is called focal point. Convergence of light leads to constructive interference and thus the intensity of light increases. This is the central concept of the light microscope and its capacity to improve image quality. The higher the intensity of light collected, the sharper is the brightness of the image. As you can see in figure 20 that the sharpness of the pinhole camera image (figure 20A) was significantly improved when I placed an eyepiece lens into the pinhole (figure 20B). Interestingly, the image formed in the pinhole camera, eye and the convex lens are real images. Real images can be projected onto a screen whereas virtual image can only be seen, not projected onto the screen.

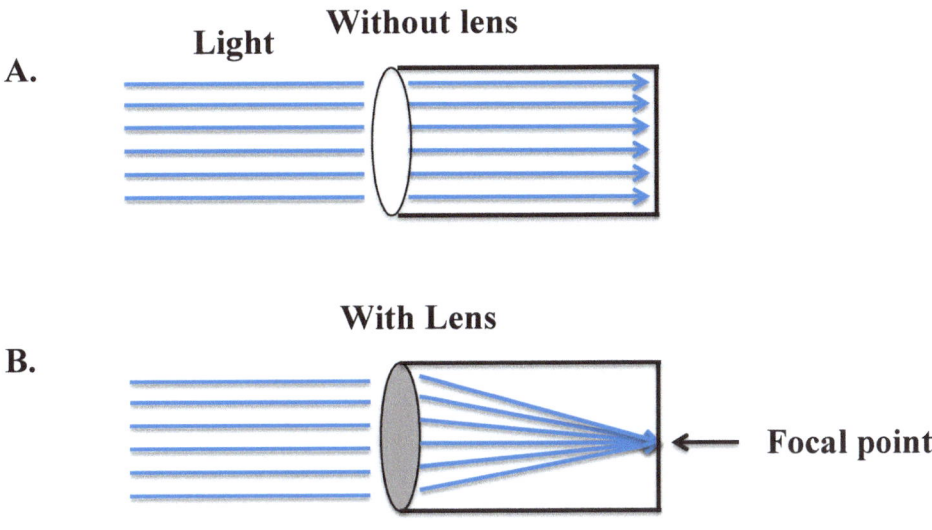

Figure 19

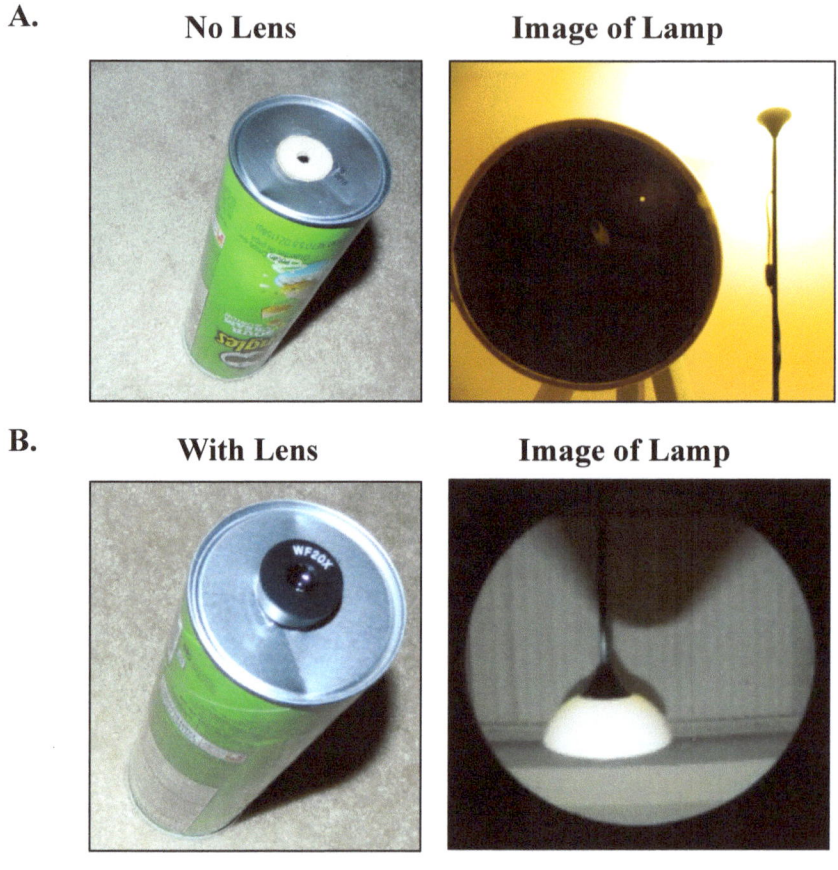

Figure 20

"By the help of microscopes, there is nothing so small, as to escape our inquiry; hence there is a new visible world discovered to the understanding".
                                                            -Robert Hooke

Microscope

Hardly three hundred years ago people had no idea that there are microbes which exist in this world and they can cause diseases. When people fell sick, they used to believe that it was because of change in weather or some poisoning in the blood. Even Geroge Washington (USA's first president) died of blood-letting, a procedure of removing blood to cure fever, believing that fever is caused by poisoning of blood. Thanks to the discovery of microbes (Bacteria) by Dutch Scientist Antonie van Leeuwenhoek, who observed bacteria under his primitive microscope in 1683, the entire field of microbes was unraveled since then. Subsequent works by Louis Pasteur (spoilage of milk by germ theory) and Robert Koch (Koch's postulate of infectious agent) established that diseases can be caused by microbes. Nonetheless, microbes are also helpful for humans and other living organisms as well as to nature. Our body has more number of microorganisms than the number of our own body cells. Gut microbiota has bazillion number of microbes which play important roles starting from digestion of the food we eat to the maintenance of our immune system. Microbes are the small organisms which cannot be seen with the naked human eye but can be seen with the help of a microscope. As the name suggests the microbes are of ~a micron size or less which can only be seen when magnified 100-1000 fold in the microscope. Viruses are even smaller in size (~100 nanometers) that can be seen by a very high to superresolution microscopy. Leeuwenhoek's microscope produced ~300 fold magnified image of microbes. In the Greek language, the word 'micro' means 'small' and 'scope' means 'to see'.

The invention of the microscope occurred around the same time as the telescope in the sixteenth century. Early microscopes (hitherto Occhilolino or little eye) by Galileo Galilei and Robert Hooke pioneered the field of microscopy. Robert Hooke who also helped develop the wave theory of light (as discussed in the previous chapter) first discovered "Cell" as a small unit of the body. He coined the term "Cell" (plant cell) as a structure resembling a honeycomb cell while looking under the microscope. Early microscopes were like a simple magnifying glass which could hardly help in seeing a bacteria or a cell. As an example, in figure 21, I have shown the image of yogurt bacteria (Streptococci- filament like bacteria) and cells from my mouth (buccal swab) captured with my home microscope- AMScope. These days, very advanced microscopes have been developed which can help us see things thousand times smaller than a bacteria. I will discuss here the basic concepts of only light microscopes which use light to produce the image, unlike the electron microscopes which use a beam of electrons or force probe microscopes which utilize atomic force to generate the image.

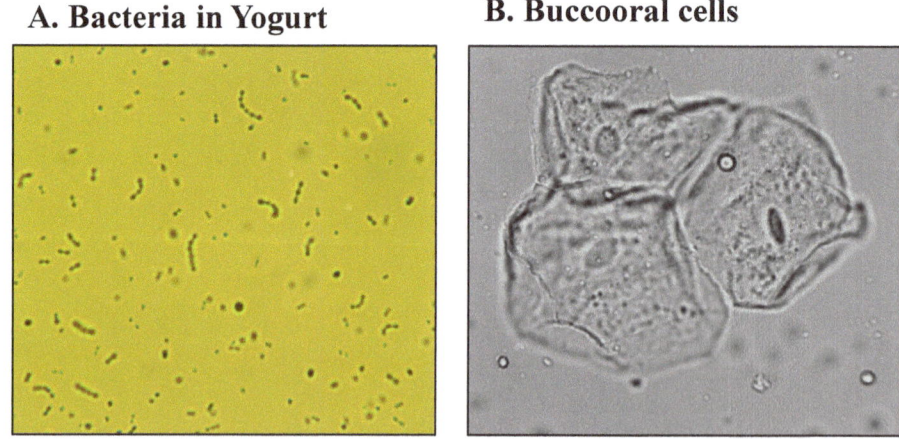

Figure 21

A microscope is a simple tube device made up of lenses to produce magnified image of the object. Microscopes in early days (like Robert Hooke's microscopes) were like simple magnifying device. A simple microscope is a microscope with a single lens. The single lens acts like a magnifying glass. I have demonstrated a simple microscope in figure 22. I assembled this wooden microscope using a 20X lens which is good enough to see small things like the compound eye of a cockroach. A compound microscope has at least two lenses. Figure 23A depicts the description of a compound microscope (AMScope). A microscope may look complicated from outside, but it is based on the simple concept of the focussing light inside (figure 23B). Three different lenses mainly govern image formation in the microscope- the condenser lens, the objective lens and the eyepiece lens. These lenses are named as per their position- the condenser lens is close to the light source, the objective lens is close to the object and the eyepiece lens is close to the eye (figure 23B).

Simple Microscope

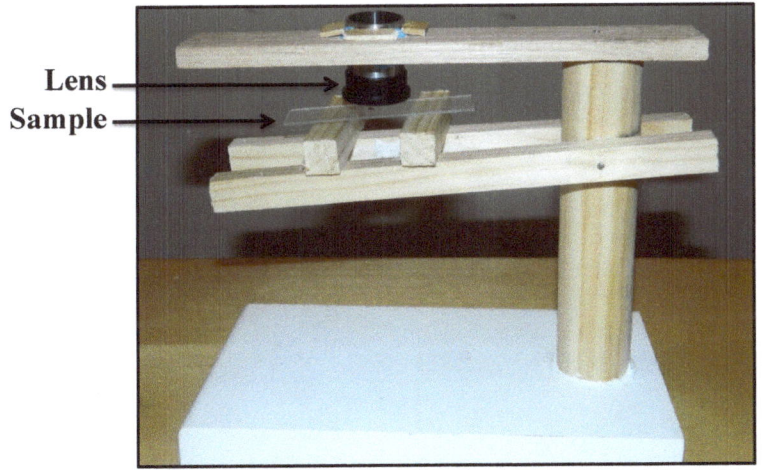

Figure 22

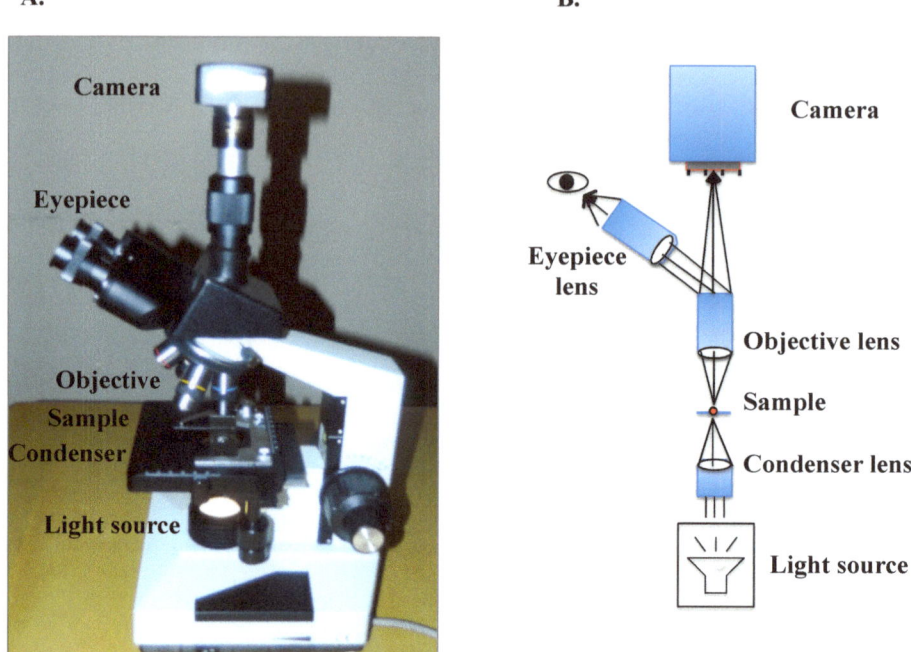

Figure 23

Let me now describe the light path inside a compound microscope. In order to visualize anything, we need illumination of the object by the light source. The light from light source is collected and focussed into the sample by the condenser lens. After passing through the sample, light is collected by the objective lens which forms a real inverted image of the sample at the image plane behind the objective. The image is then projected either into the eye through the eyepiece lens or to the camera chips through additional lenses. By the way, when we look at the object with the microscope, we actually do not see the object. Instead, we see the image of the object in the intermediate plane (inside the microscope) on the rear side of the objective lens.

Magnification

Magnification in the microscope is achieved mainly by the objective lens and then additional magnification by the eyepiece. The magnification can be measured from the relative position of the object and image to the lens which is basically, M=di/do (where M is magnification, di is the distance of the image from the lens and do is the distance of the object from the lens (figure 25A). If the microscope has an objective lens of 100X (means the image it forms is of 100 times zoomed) and an eyepiece of 20X, then the final image formed is of 100X20=2000 times magnified. This is how the microscope helps us to see very small things that cannot be seen by the naked human eye.

A human eye can see things as small as ~0.1mm, but with the help of a general microscope, we can see 0.1x2000 fold magnification, which can be as small as 200 nm which is the size of about a large virus or a very very small bacteria. I have shown in figure 24 how the objective lenses of different magnification increase the image size of seaweed cells (seaweeds samples collected from Sandy Point Beach, MD, USA). You can see that as the magnification of objective lens increases, the size of cells increases. Interestingly, higher magnification lenses are longer (thus it forms the image of the sample from the nearer distance) and have the smaller apertures (opening at the end of the objective lens). As higher magnification lens positions closer to the sample, it collects light from a smaller region of the sample, allowing more light per area of the object which means better resolution and bigger image size. It simply fits into the above equation, M=di/do. If the distance of the object from the lens (do) decreases, then magnification (M) increases.

Magnified Images of Seaweed cells

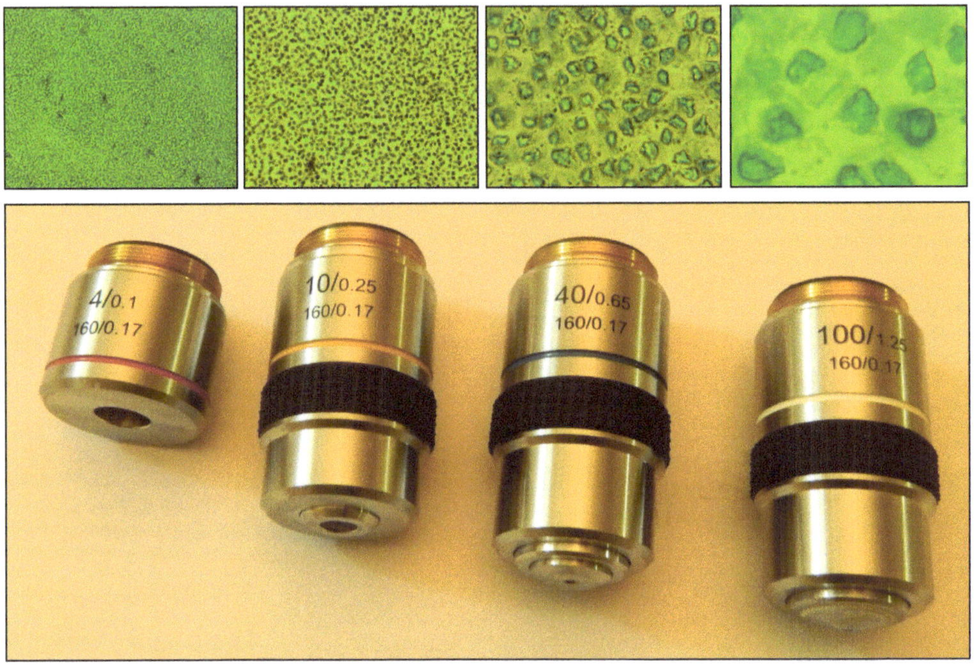

Figure 24

I will now explain how an image is formed in the microscope. If we understand how an image is formed, then we will understand how small we can see and what is the limitation of the microscope. The next chapter will deal with the most of the basic concepts of points in object and image.

"There is no absolute point anywhere in the world. Just like a point is made up of many points, an image is also made up of many images".

In order to understand the concept of image formation through lens, we usually refer to ray diagram. Figure 25A shows one of such diagram of the real inverted image of an object on the other side of a biconvex lens. Does it really help us to understand how rays unite to make image? Let us discuss it in more detail. An object is not a single point source of light, rather each point in the object acts like a point source. Thus an object can be thought of as various points that make up the object. Light coming from each point of the object forms a point on the image (figure 27). As an example when we image a cell, the image we get is generated from the light coming from different points of the cell (figure 27). Each point can be called a point source of light. There is no absolute point in the universe, but for the convenience of understanding, we can think a point in an object as a small unit of light source. Figure 25B shows a point in the object (red dot) which emits light. The lens then collects the light from the point source. After passing through the lens, light is merged to a point to form an image point. In the previous section we have already discussed why light merges after possing through biconvex lens.

For the simplicity of understanding the direction of propagation, I drew light as rays in figure 25B, but if we think of light as a wave, the light from the point source in the object actually comes towards the lens as waves (figure 25C). Now we understand how each point in the object behaves. The same way we can think about other points of the object. In figure 25D, I have shown just three different points (in blue, green and red dots) to explain that different points from

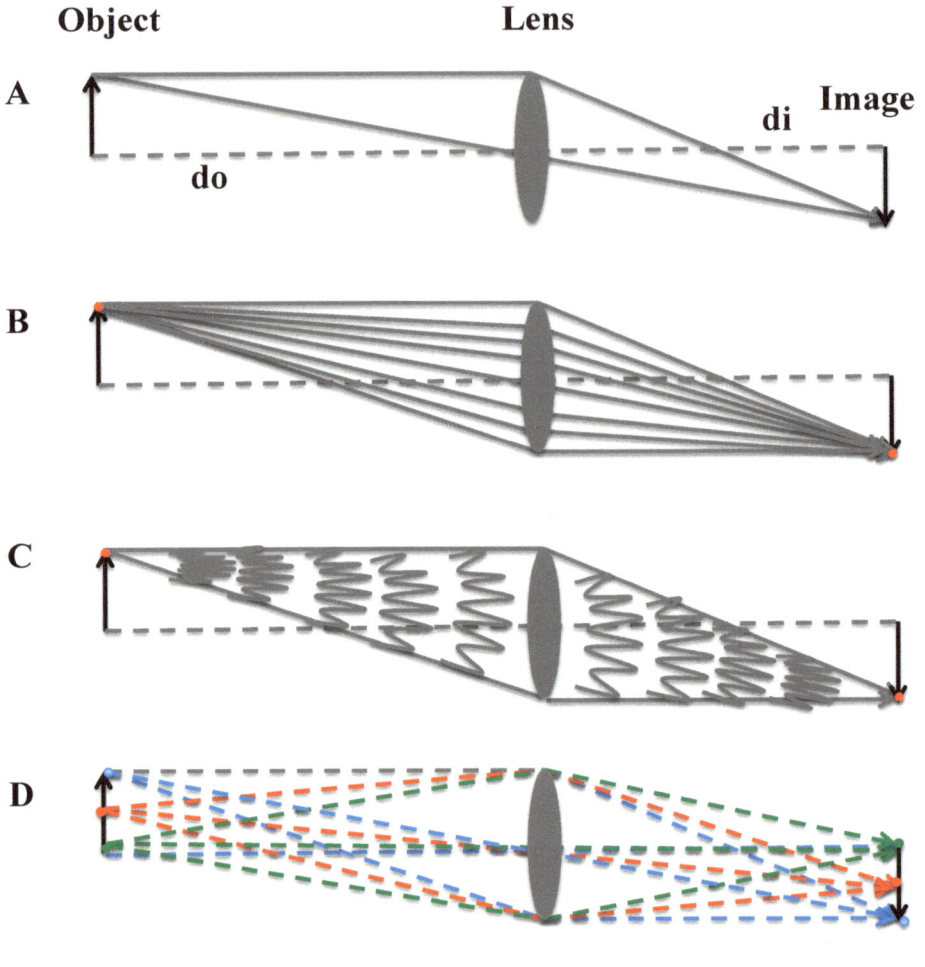

Figure 25

the object spread out light which, after passing through the lens, merges at three different points on the image plane (figure 25D). Thus an image point is generated by constructive interference of light after passing through the lens, which generates a brighter spot. Again, a point in the object not only spread light towards the lens. In other words, the lens is not capable of fully collecting all the light generated by the object points (figure 26A). A point source in object spread out light all around just like the Sun or an electric bulb. Let's think that the Sun is a point source of light and spreads in all directions in spherical wave fronts (figure 6). However the Sun is not a point source, rather each point in the sun is a point source. Like wise, a point in the object emits light in all direction, the lens collects only a fraction of it (figure 26A). Moreover, the light after passing through the lens not only merge and stop right at the point on the image. Rather the light diverges again after the point (in the image) further below if no additional blocking or absorbing media exist beyond that (figure 26B). Just for the simplicity of understanding, it is shown that light from the object forms a point in the image.

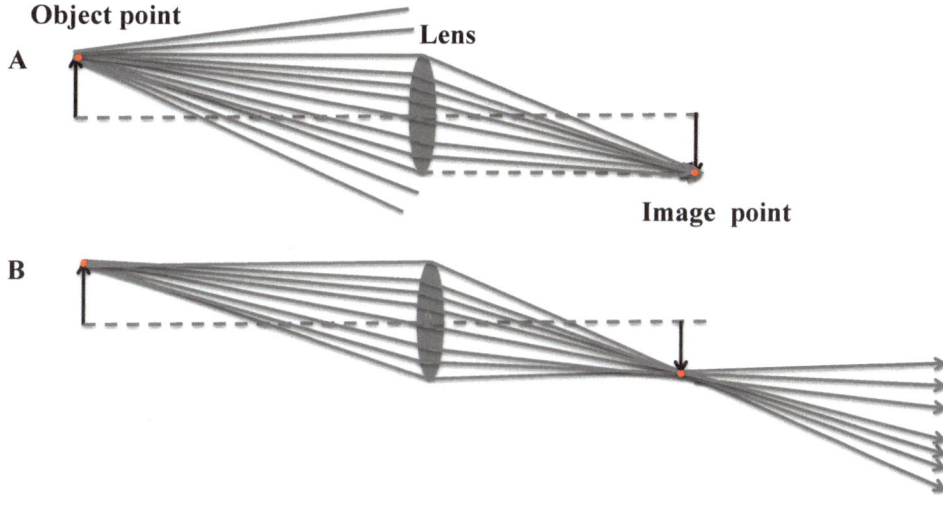

Figure 26

A. Image of a cell	B. Points on the image
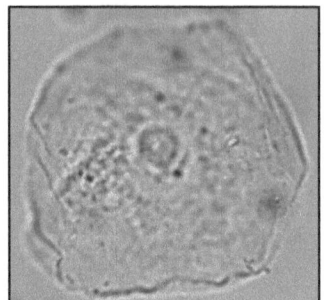	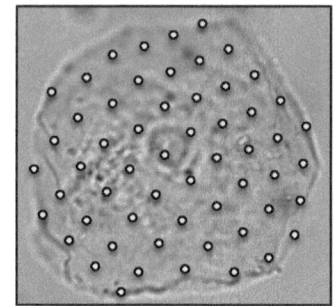

Figure 27

As mentioned above, an image can be constructed from various points originated from the object (figure 27). A similar comparison can be drawn from living room light (figure 28). Here we can think of each living room light bulb as a point source. When I projected the image of five light bulbs with a magnifying glass onto the wall, I observed five bright spots/points corresponding to each bulb (figure 28A). If we think the points on the wall as the image of the whole set of light bulbs then each point on the wall originates from each point on the object (each bulb). If we further adjust the focus, each point on the wall actually looks like the inverted image of the bulb (figure 28C). Similarly, when I took a pinhole camera (above mentioned potato crisps container with a hole), I saw five bright points as the image of bulbs, one point corresponds to each bulb (figure 28B). This experiment directly tells us that an object can be thought of as a light emitting point.

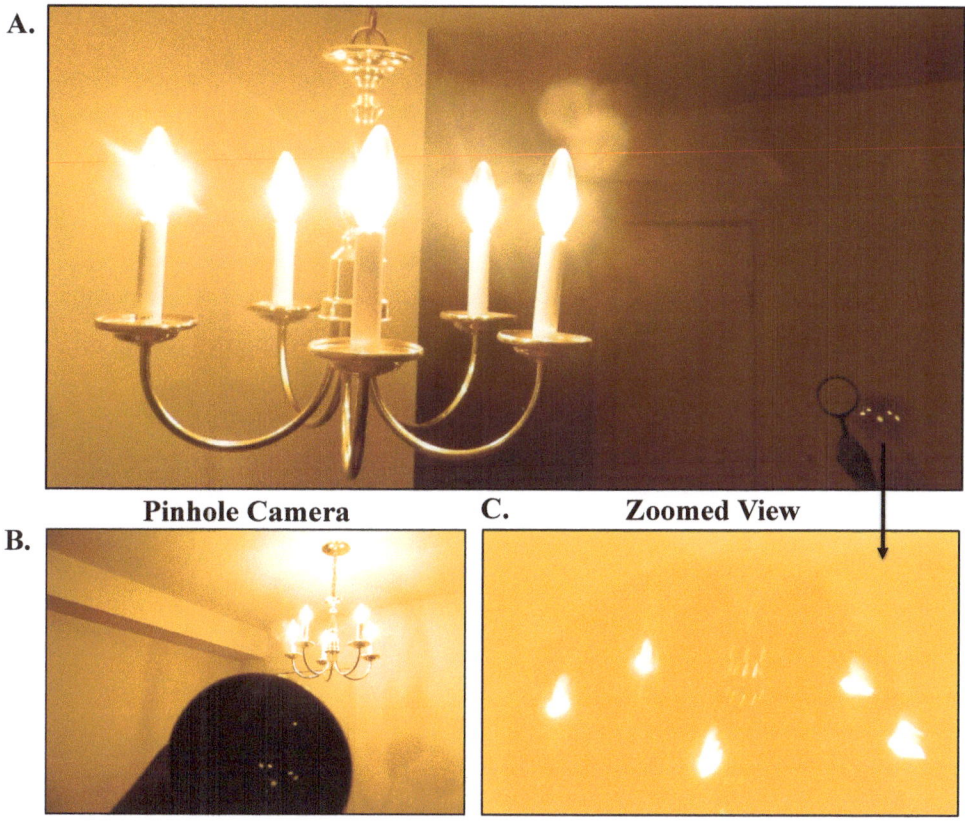

Figure 28

Focal point vs. Image point

Sometimes it becomes confusing to understand the difference between the focal point and the image point. Why does the lens not form the image on the focal point? The focal point of the lens is the position where parallel lights (either coming from infinity or a nearby object) converge together (figure 29A and B). Let's imagine three points on the object from where these parallel lights (blue, green and red arrows) emerge (figure 29B). After passing through the convex lens, they converge at the focal point. In order to form an image, all the light rays (light cone) coming out from each point of the object should converge at a point behind the lens (figure 29C). For a simpler understanding, I have shown all the lights coming from the green point as a cone of green light in figure 29D. When the image plane is fixed, as in the camera or our eye, we adjust the distance between the lens and the object to bring the image into focus. This is exactly what the focus adjuster of a microscope does. The focal length (f), the distance of the object from the lens (u) and distance of the image from the lens (v) are related by the following formula-

$$1/f = 1/u + 1/v.$$

Hence the focal length of a lens can be calculated from the above formula by measuring the distance of the object and the image from the lens. A biconvex lens has the focal length on either side of the lens- the one on the front side of the lens (side facing the object) is called the front focal point and the one on the back side of the lens (facing the image) is called the back focal plane. In a microscope the image of the object (sample) is formed behind the back focal plane of the objective lens which we see through the microscope's eyepiece.

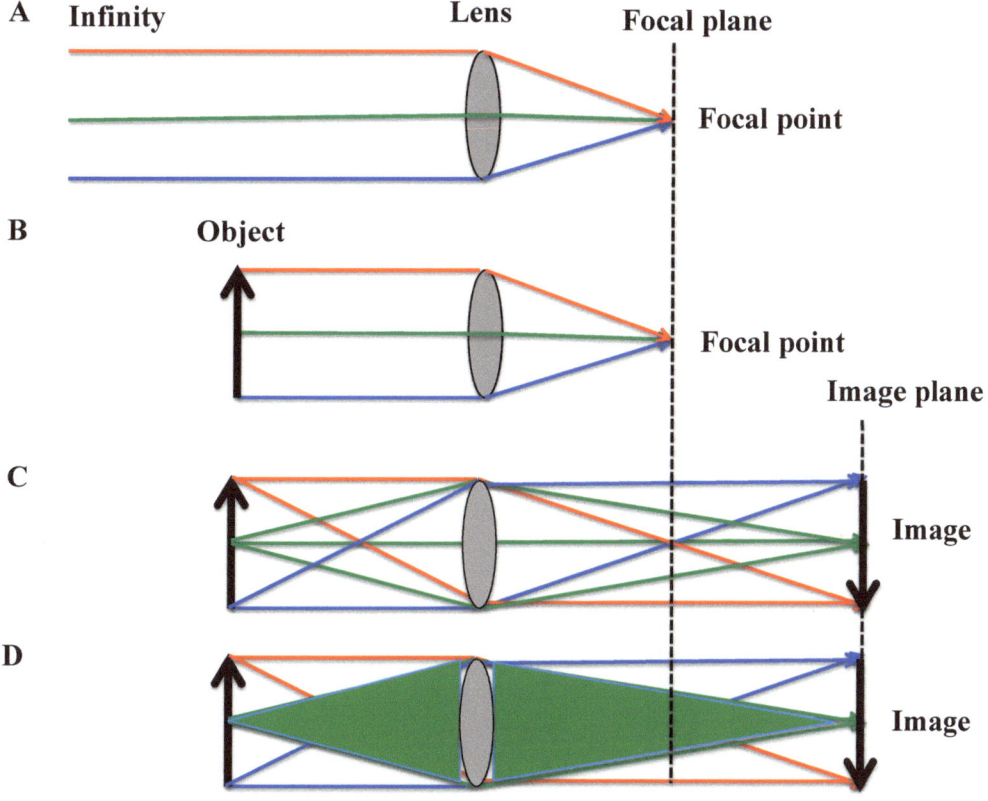

Figure 29

Imagining each object point as a light source will help us understand how the objective lens of the microscope forms the image with proper resolution. This will also help us understand the limit of resolution in light microscopes. A point source in an object emits light all over (figure 30, left panel). In most of the conventional light microscopes, the light is collected by a single objective lens (figure 30, middle panel). However, a single objective lens can collect only a fraction of the emitted light. The higher the amount of light collected by the objective lens, the better is the resolution of the image. We will discuss this concept in the next chapter. To collect more light from each point of the object, 4Pi microscope utilizes two objective lenses positioned on either side of the object (figure 30, right panel). One objective lens can theoretically collect light up to 180° (2π or 2Pi) angle. Thus, with the help of two objective lenses, the microscope can collect light up to 4π (4Pi) angle. However, this (360°) light collection is not practically possible. Still, 4Pi microscope improves the resolution of the image (up to 100 nm) compared to the conventional light microscopes.

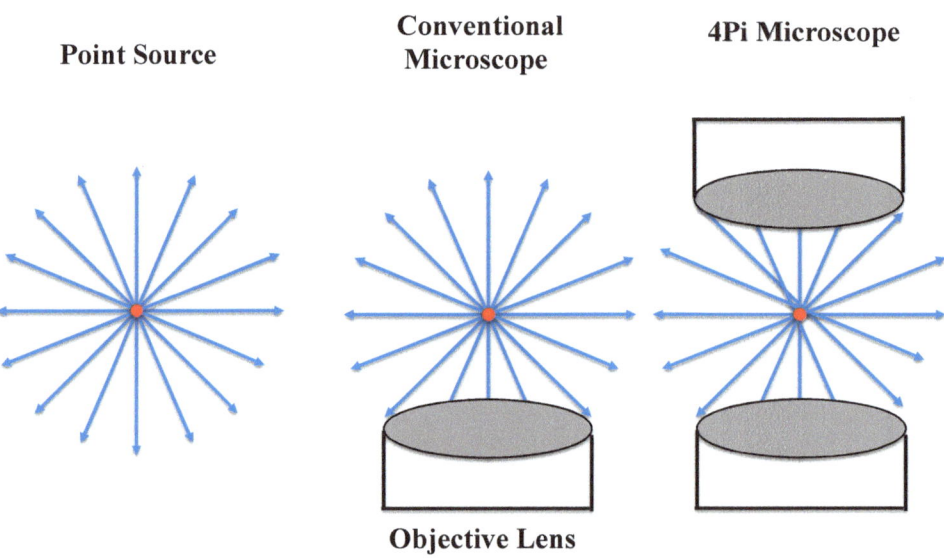

Figure 30

Numerical Aperture

The most important feature of a microscope lens is its numerical aperture or NA in abbreviation. 'Numerical' refers to 'number' and 'aperture' refers to 'the opening of the lens'. NA is the numerical definition of the aperture or the opening of the lens (figure 31A). The lens has the ability to collect light from each point of the object up to a certain area. The cone of light received by the objective lens governs the quality of image formation. Numerical aperture is the product of refractive index (n) of the medium and the sine value of semi-angle (θ) of the cone of light received by the objective lens.

$$NA = n \cdot \sin\theta.$$

The higher the NA, the better is the light gathering capacity of the lens and thus the better the resolution of the image formed by the lens. The refractive index of the intervening medium between the objective lens and the sample also regulates the amount of light entering the eye. Therefore, to increase the axial resolution, immersion oil of higher refractive index (~1.5) is used in many cases. As a medium of higher refractive index bends light more towards the normal (towards the lens here) (figure 31B), more light enters the lens when oil is used compared to air or water (figure 31C). Therefore an objective lens with higher NA requires immersion oil of high RI to produce better resolution image.

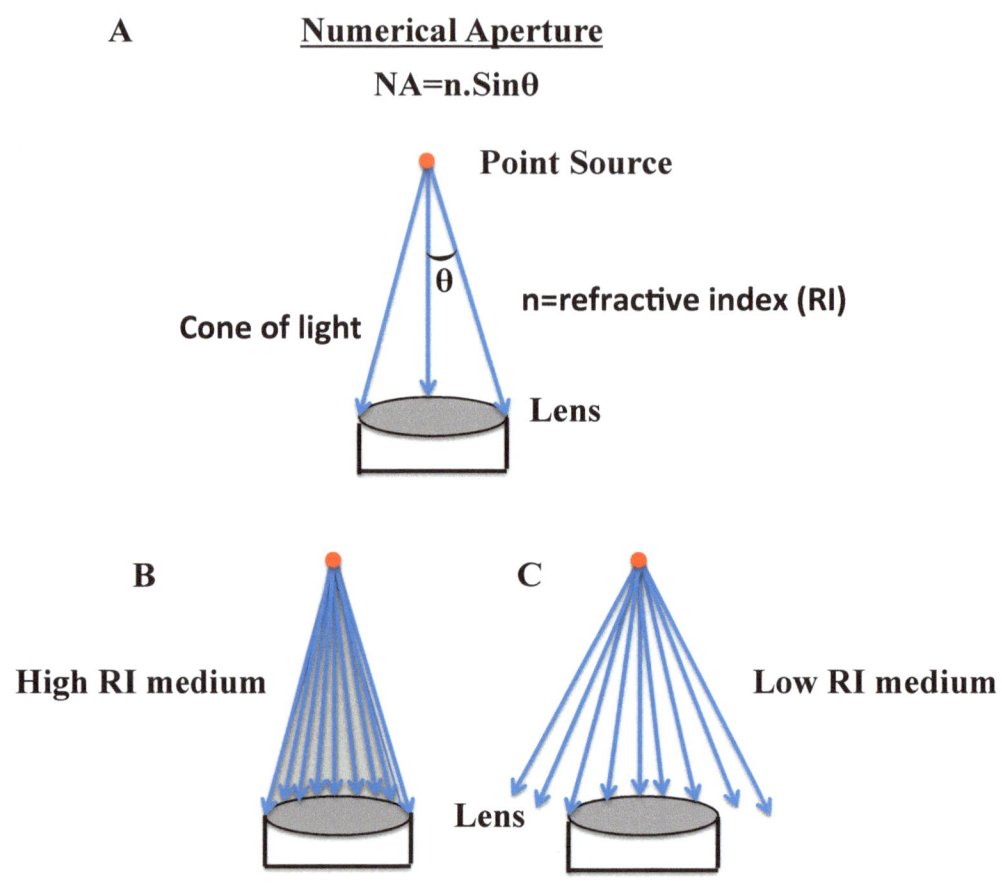

Figure 31

General microscopy

There are various types of microscopy techniques developed depending upon the need of study goals. I will discuss some of the commonly used microscopies which are within the scope of this book. I will focus on a few techniques in microscopy that describe how different properties of light such as wave or particle nature, fluorecnence, interference and reflection govern special microscopy applications.

Bright-field microscopy

Simple low resolution microscopes like a bright-field microscope collects light transmitted through the samples to generate image (figure 32 left panel). Here, the image of the object looks darker than the surrounding field, thus the field is bright. Contrasts in different parts of the objects arise due to attenuation of light passing through different spots of the sample with different densities. The limitation of this technique is that too many details of the object cannot be obtained with bright-field image.

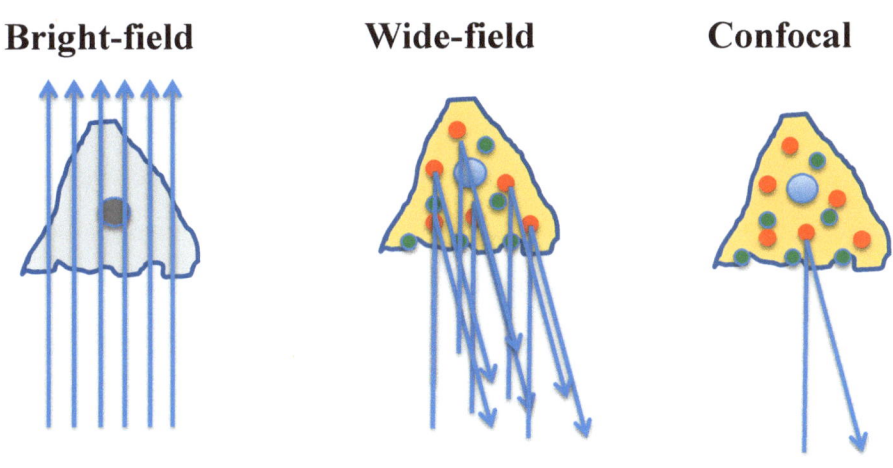

Figure 32

Fluorescence

Fluorescence is the phenomenon of the release of a photon of longer wavelength when electrons of fluorophore atoms relax from higher energy singlet state to ground state. Fluorophores absorb light of specific wavelength. Electrons in the atoms are associated with different vibrational energy levels at the ground state. When the atoms of fluorophores absorb light of certain wavelength they move to singlet states like S1, S2 etc. (figure 33). Upon relaxation, the electrons loose some energy in the form of non-radiative loss like heat to come to lowest singlet state vibrational level (S1). The rest amount of energy is released as a photon of higher wavelength when the electron finally comes down to the ground state (figure 33). This process of fluorescence occurs in one billionth of a second. Due to loss of energy during this process of absorption and emission, the number of photons emitted is less, compared to the number of photons absorbed. This is called the quantum yield (number of photons absorbed/number of photons emitted). Fluorophores' quality or brightness is determined based on the quantum yield (QE). Maximum QE possible for a fluorophore is 1. The phenomenon of fluorescence makes it possible to gather specific signal (red dots in figure 32 middle panel) and exclude non-fluorescent signal of the object (green dots in figure 32 middle panel). We can also label proteins with multiple fluorophores and image multiple markers in the same cell.

Fluorescence

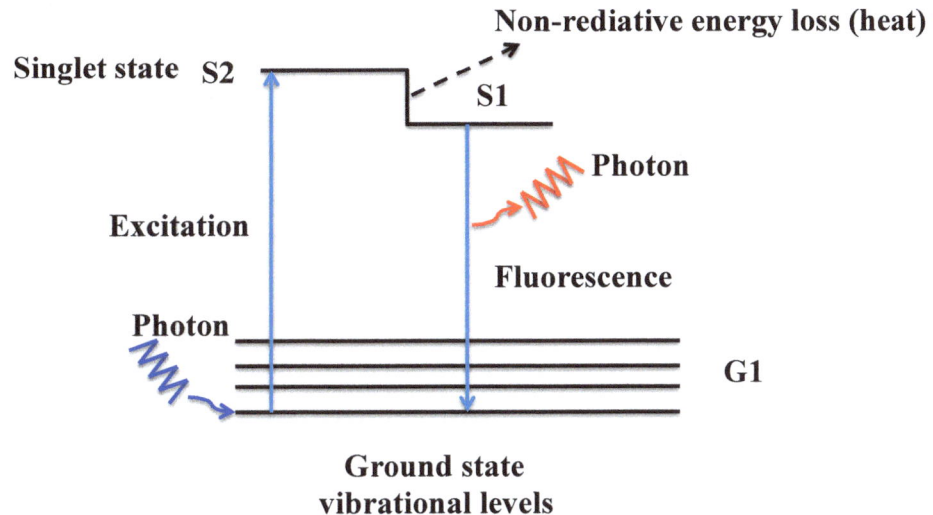

Figure 33

We can notice fluorescence while looking at jelly fish, Glofish or gemstones. Huge application of fluorescence was achieved when Japanese marine biologist Osamu Shimomura discovered a naturally occurring fluorophore called Green Fluorescent Protein (GFP) from jellyfish *Aequorea* in the1960's for which he was awarded the Nobel Prize in 2008. Discovery of GFP revolutionalised the field of microscopy as it made possible to link fluorescence to image proteins or organelles inside the live cell, tissues or even the whole animal. Chemical fluorophores like Alexa or FITC dyes are commonly used in the biomedical research to image specific proteins in the cells.

Fluorescence microscopy is based on labeling specific set of molecules with fluorophores which can be selectively excited with light of specific wavelength. This makes it possible to exclude non-fluorescent signal of the object. Therefore fluorescence microscopy improves the resolution of the image as the molecules of interest in the object are fluorescenet (figure 32 middle and right panel with green and red dots).

Wide-field epifluorescence microscopy

One example of commonly used fluorescence microscopy is wide-field epifluorescence microscopy (figure 32, middle panel). Wide-field implies that the whole field of view is illuminated with light and thus the image of the whole field of view is captured by the camera. The word "epi" means same. Epifluorescence implies that the excitation light to sample and the emission (fluorescence) light from sample goes through the sam path of the objective lens. In this type of microscopy, the sample is illuminated with light of a specific wavelength which the fluorophores in the object (figure 32 middle panel, red dots) absorb and emit fluorescence. By using wavelength specific excitation and emission filters we can selectively image specific sets of fluorescent molecules (red dots) and discard other non-specific molecules (figure 32 middle panel, green dots). The disadvantage of this technique is that it does not exclude out-of-focus light which blurs actual signal from the fluorophores of focus (red dots). Nevertheless, wide-field microscopy is widely used for its convenience, faster image acquisition, less photobleaching and cheaper price compared to other microscopy techniques.

Confocal microscopy

The confocal microscope is very popular in biomedical research as it significantly improves the signal to noise ratio by using pinhole aperture. The pinhole aperture is optically connected to (or related to) or in other words, it is in the conjugate plane to the front of the detector. Thus, a pinhole aperture blocks out-of-focus light and allows fluorescent light only from the point of focus in the object (figure 32 right panel and figure 34). By sequentially scanning every point in the object with the help of a laser point scanner or a disk full of pinholes (Nipkow/Yokagawa disk), the confocal microscope generates an image with enhanced resolution.

Principle of Confocal Microscopy

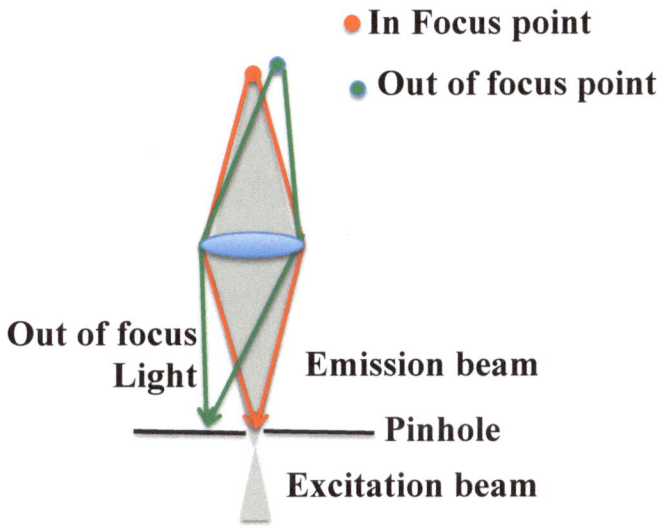

Figure 34

Total internal reflection fluorescence (TIRF) microscopy

Total internal reflection fluorescence (TIRF) microscopy is one of the wide-field microscopy techniques which enables optical sectioning. Optical sectioning refers to the process by which microscope designed with special optics generate high-resolution image of focal planes (specific section) within the whole stack of tissue/cells. Let me explain what total internal reflection is. When light enters from a medium of high refractive index to low RI one, the direction of light changes depending on the angle of incidence (figure 35 A-D). When light is incident along the normal, it enters undeviated straight into the 2nd medium (figure 35A). When light is incident at an angle less than the critical angle, it bends away from the normal (figure 35B). When light is incident at the critical angle, it bends along the interface (figure 35C). When light is incident at an angle greater than the critical angle, the light undergoes total internal reflection (figure 35D). As the light behaves total internal reflection, it does not bend right away from the interface, rather it gradually fades, thus the nearfield evanescent wave can penetrate up to 100 nm into the second medium (figure 35E). This evanescent wave gradually fades away. This special penetration depth allows TIRF microscope to do optical sectioning by imaging fluorophores of the object within 100 nm from glass slide (interface), thus excluding out-of-focus fluorescence light coming from the points deep inside the sample (figure 35F and G). TIRF microscope uses the laser as incident light to generate sufficient intensity of specific wavelength light needed to excite fluorophore. This technique is commonly used to image proteins close to the plasma membrane of the cell.

Principle of Total Internal Reflection

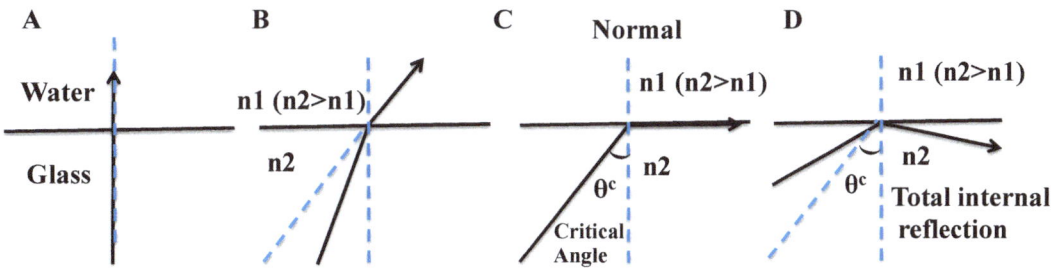

Total Internal Reflection fluorescence (TIRF) microscopy

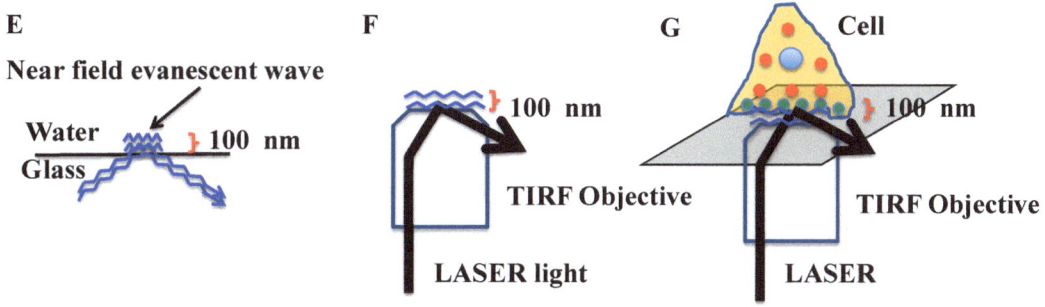

Figure 35

Interference reflection microscopy

Interference reflection microscopy (IRM) is based on the principle of destructive interference of light. When an object (let's say a spreading cell) comes in contact with the glass slide, light is reflected from various parts. Most of the incident light is scattered by the sample, but around 1% of incident light is reflected back into the same medium (glass) from where it first came. A cell cytoplasm has the similar refractive index (n~1.33) as the surrounding aqueous medium (water) but less compared to the plasma membrane (n~1.48) or glass (n=1.5). As the cell membrane spreads over the glass slide, lights are reflected from the various parts of the membrane (figure 36). Because of the difference in refractive index, light reflected from a glass surface and the cell membranes will have phase difference and thus results in destructive interference. As the cell membrane approaches closer to the glass slide, the phase difference of light reflected from cell membrane with that from the glass surface becomes more (up to half wavelength). This results in destructive interference and the darker image of cell whereas the membrane of the cell farther away from the glass slide produces a brighter image (figure 36 bottom panel). This interference based microscopy is very helpful to study the behavior of cell adhesion and migration.

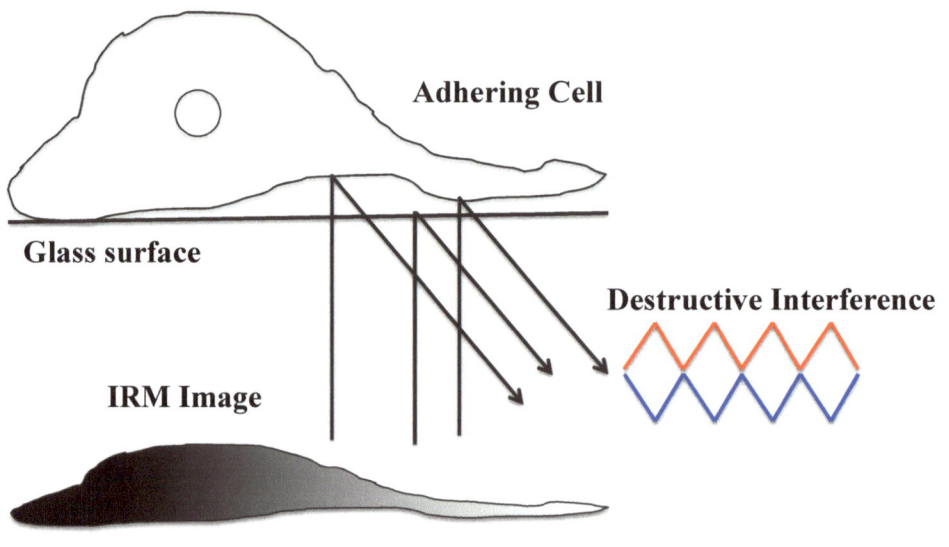

Figure 36

"Light helps us to see, at the same time light also hinders us from seeing".

How small can we see?

The microscope is one of the noblest inventions of human beings. Ever since microscopes were developed, scientists have been curious to improve the power of the microscope. The million-dollar question is "How small can we see? Until the end of the twentieth century, the resolution of microscopes was limited by the very fundamental property of light- "diffraction". However, in the first decade of 21st-century scientists developed super-resolution microscopy techniques that counteract the diffraction of light and can produce image beyond the diffraction barrier. Stefan W Hell was awarded the Nobel Prize in 2014 for developing one such super-resolution technique called STED microscopy. I will describe the super-resolution technique later in this section.

First of all, let me describe what regulates the resolution. The simple rule is that the higher the amount of light collected per unit area (or points) of the object, the better is the resolution of the image. The resolving power of any optical instrument (like eye or microscope) is the capacity to resolve or distinguish between two closest points possible. One of the approaches to achieve better resolution is to capture image closer to the object. As the objective lens approaches closer to the point source in the object, more light is collected (figure 37). As any point sources can emit light in spherical wave fronts, the density of photons decreases as we move farther away from the object point. Thus an objective lens situated closer to the object will generate higher resolution image than the one situated farther. This is how higher NA and higher magnification objective lenses like 60X, 100X and 150X are longer and positioned closer to the sample (figure 24).

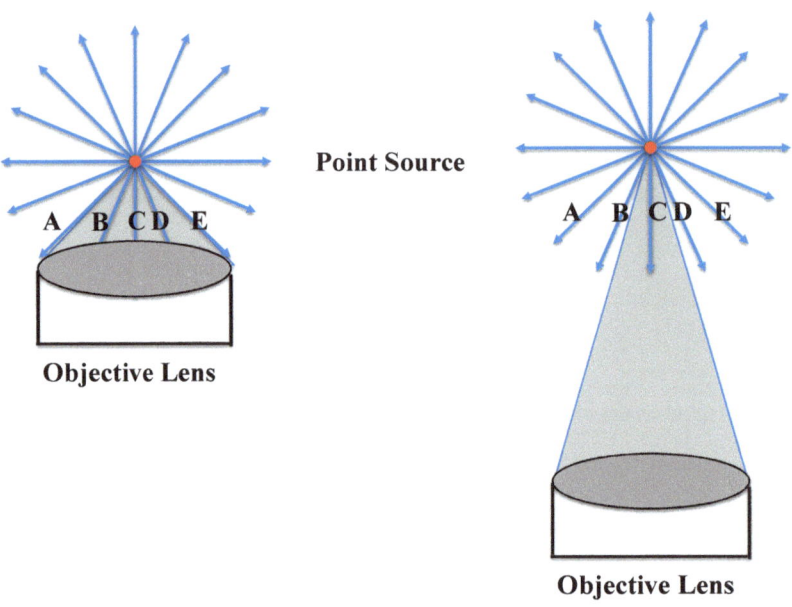

Figure 37

One common example to explain the above phenomenon is taking the image of any object closer to us. In figure 35, I have shown the image of stop lights (multiple red bulbs) on the backside of a car taken from far or near to the car. When imaged from far, the multiple lights look merged (convoluted) and appear as a single light bulb (figure 38 left). As I captured the image closer to the car, the different bulbs are seen clearly distinct from each other (figure 38 right). This is exactly what happens when we image objects in the microscope with the different objective lenses. Just as the light waves from each bulb generate diffraction patterns, which is the main reason for obscure image formation, the lights coming out from nearby points in the object cause diffraction patterns.

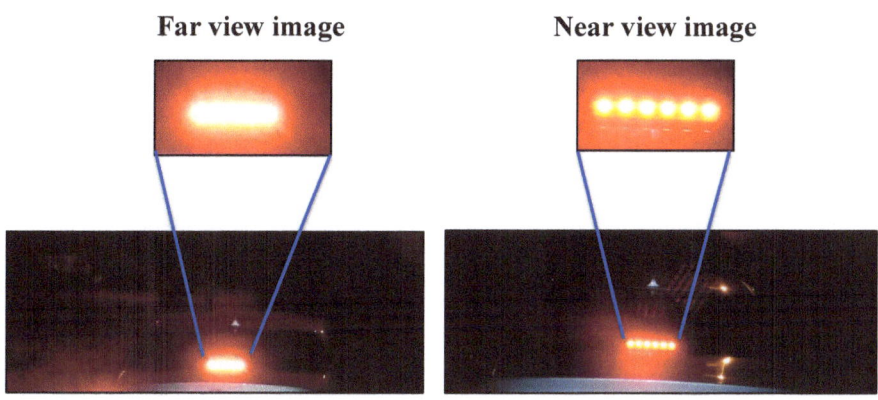

Figure 38

Ideally, in the absence of diffraction, a point in the object should generate a point of similar size and intensity in the image (figure 39A). The intensity of the point (shown at the bottom of the figure) should be limited to a very small region in X and Y planes (as shown) and also in the Z direction (not shown). However, in reality, the image generated by the conventional microscopes is made up of points whose intensity is diffused because of diffraction as shown in figure 39B. Let me explain how diffraction occurs in the microscope. The aperture of an objective lens is basically a hole. When the light comes out of this aperture of the objective lens, lights bend away to generate the diffraction pattern at the image plane. In addition to the objective lens, diffraction can arise from the samples when light travels from the point source towards the lens. The diffraction limited spot in the image shows intensity spread over a distance in the XY plane as shown in figure 39B right panel.

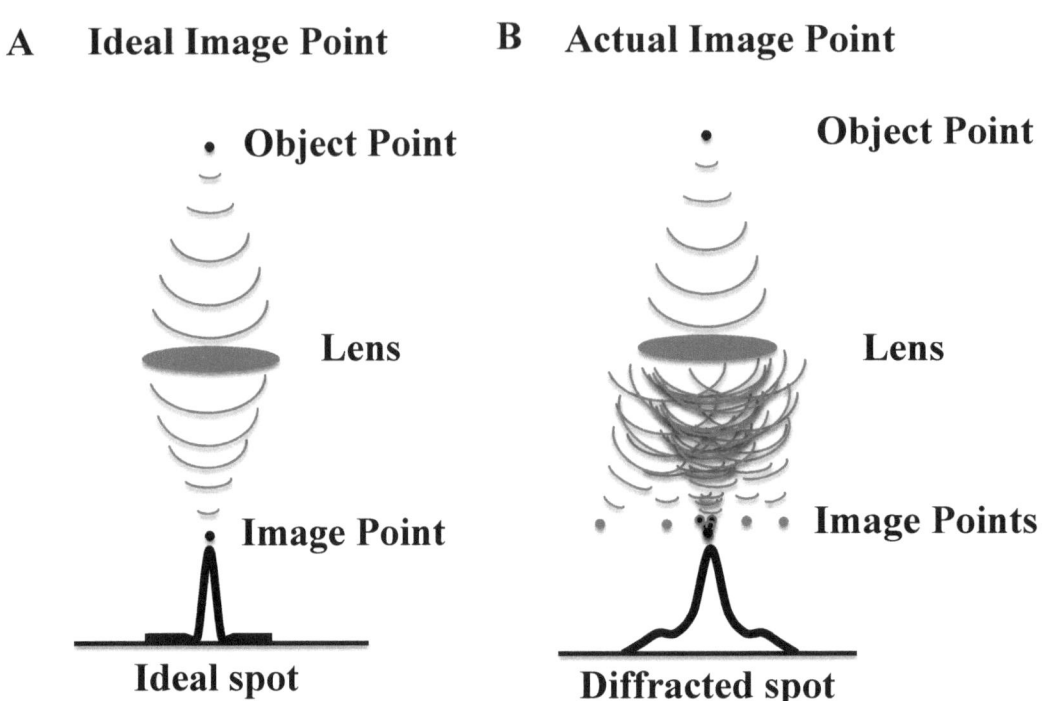

Figure 39

We can notice the phenomenon of diffraction everywhere we see light coming out of apertures, be it a street light or car head light or stop lights in the street. I have shown one simple example in figure 40, where diffraction pattern appears clearly when I projected laser light (commonly used laser pointer) to the ceiling (figure 40A). The center of the spot appears to be a tight bright spot (undiffracted) which is surrounded by rings of diffracted laser light. This is also called Airy disc pattern. After capturing the image of the laser spot, I did a line scan of the image by Image J software (as shown in figure 40B). The line scan graph shows that intensity is spread over XY plane with the highest intensity in the center of the spot. By numbering the peaks of the point spread, we can label the central brighter peak as 0, the subsequent small ones as 1, 2, 3 .... (figure 40B inset).

Now we know that diffraction is closely associated with propagation of light through apertures. This concept is important to understand the points where diffraction creates a problem in resolution in the microscope. As discussed previously, light in the compound microscope is focussed mainly by the condenser lens (figure 41A and figure 23). Accordingly, diffraction occurs at the object point as the condenser lens focuses light on the object (figure 41B). This diffraction is generated as the light comes out of aperture from condenser lens. The second major diffraction occurs at the image point where light is collected by the objective lens to form an image (figure 41C). This leads to diffraction limited spots in the image which is the major limiting factor for resolution of an image.

Conventional fluorescence microscopes like wide-field and confocal microscope do not utilize condenser lens. Here, the objective lens also behaves as the condenser lens to both, focus light on the sample and form image (figure 42A). As the image is produced by fluorescence light, the same objective lens can focus fluorescence light (emission light) on image after passing through filters in a different light path. In this case also diffraction occurs at both object point and image point (figure 42B).

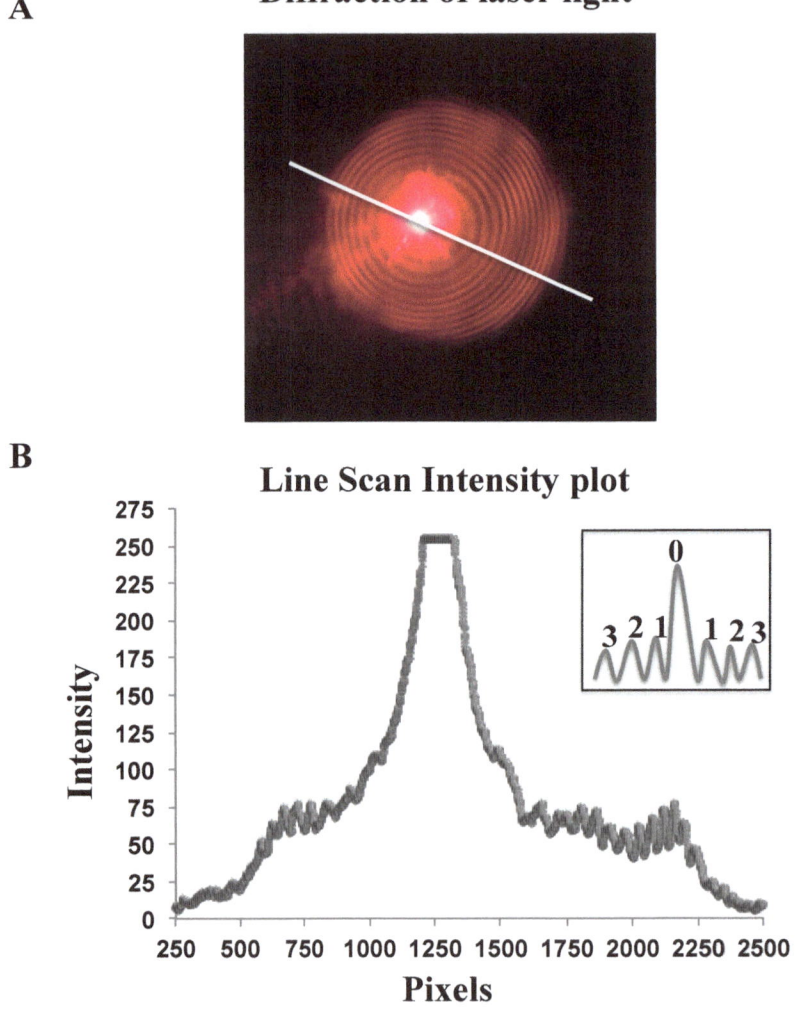

Figure 40

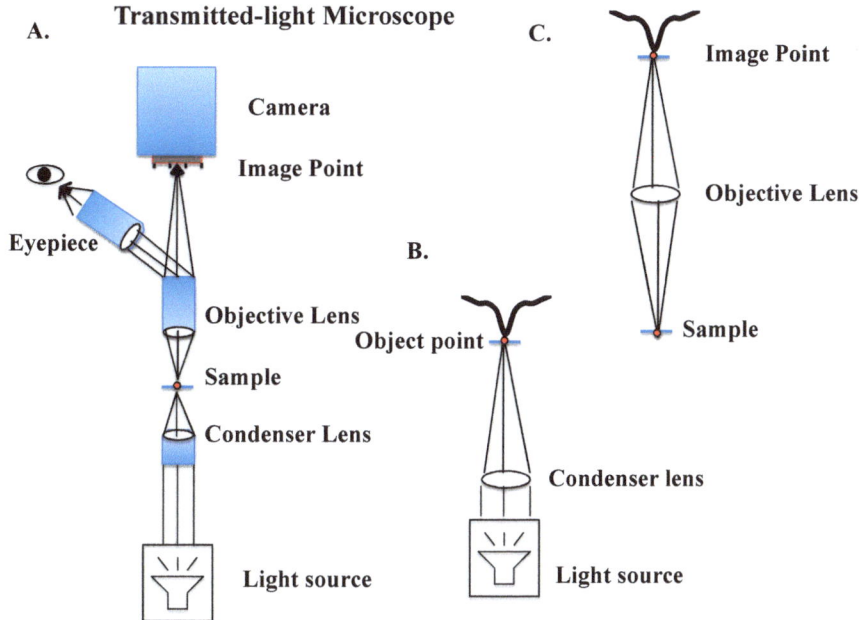

Figure 41

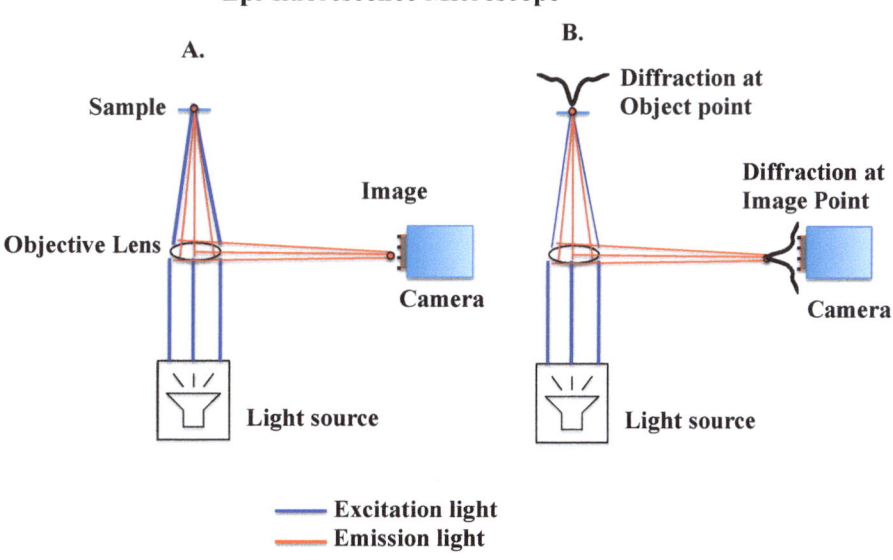

Figure 42

The limit of resolution

Considering the diffraction nature of the light, Prof. E. Abbe in 1873 derived the mathematical definition of resolving power for conventional light microscopes. The minimum distance between two points to be resolved as separate or distinguishable is called limit of resolution. Abbe defined the limit of resolution (d) to be dependent upon the wavelength of light (λ) and the numerical aperture (NA) of the lens.

$$\text{Abbe's limit, } d = \lambda/2NA$$
$$\text{Or } d = 0.5\, \lambda/NA$$

In the above equation, the 2NA refers to Numerical Aperture of the Condenser + Objective lenses. In wide-field or confocal microscopes, the objective lens also behaves as condenser lens to focus light onto the sample. Thus in these cases (where there is no separate condenser lens), it is taken as 2NA. For white light, the wavelength is an average of the entire visible spectrum (~570 nm) So, according to Abbe's limit, a conventional microscope with objective lens of NA=1, using excitation light of wavelength 500 nm, should be able to achieve the limit of resolution- 250 nm (d=500 nm/2). However, this is practically not achievable. If we image a sub-resolution bead (of ~100-200 nm or less than the wavelength of light), with excitation light of 500 nm wavelength, then the image of bead will look much bigger as shown in figure 43. If we draw a line across the image of bead and measure the intensity profile, then we will find the point spread function. The limit of resolution can be calculated from the full width at half maximum (FWHM) intensity. Exact calculation of actual point spread is possible these days because of defined pixel sizes in the digital camera. However, back in 19th century (Abbe's time), the limit of resolution was calculated by visually looking at the image through the eyepiece, which can be prone to subjective error. Even Abbe was concerned about this kind of error.

Lord Rayleigh derived a modified version of the limit of resolution equation. According to Rayleigh's criterion, the resolution (R) = **1.22 λ/NA** or simply **0.61 λ/NA.** The number 1.22 is derived from the position of the first dark circular ring surrounding the central airy disc (or the central spot of point image). Mathematically it is the first zero of the order one Bessel function of the first kind divided by π. Nevertheless, both Rayleigh's criterion and Abbe's limit are of approximately similar value.

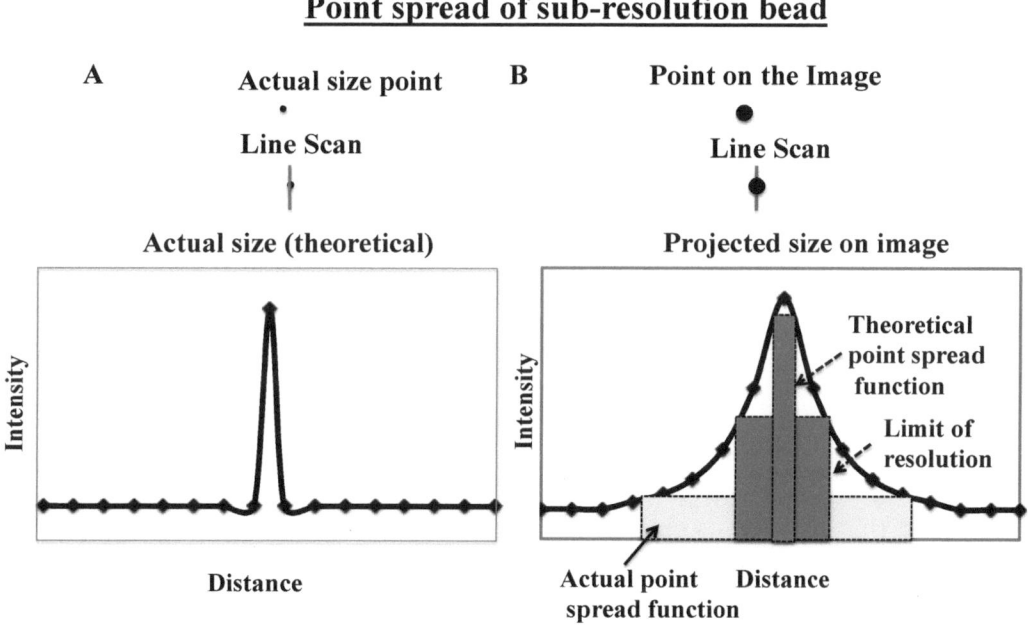

Figure 43

Super-resolution microscopy

So far, we came to know that using conventional light microscopes the smallest thing that can be seen is ~200 nm because of the limitation owing to diffraction. The diffraction nature of light propagation would not be a problem if only one molecule in the sample were fluorescent (or emitting light). We can still be able to measure the point-spread function and calculate the center of the molecule. The problem arises when there are many fluorescent molecules within the main diffraction spot of one point (or zero of the point spread, figure 40B inset). Then point spread of the entire fluorophores makes the spot even more diffuse and difficult to resolve individual fluorophores. However, there are a few approaches developed in the last decade, to counteract the diffraction barrier. These techniques called super-resolution microscopy have helped achieve the limit of resolution much smaller than Abbe's limit, even up to 1-20 nm.

There are various types of super-resolution techniques, but the description of all is beyond the scope of this book. Basically, they can be divided into two main groups. One group involves techniques like STED (stimulated emission depletion) and GSD (ground state depletion) microscopy. These techniques which selectively de-excite fluorophores around the central spot to counteract most of the diffraction barrier are called deterministic super-resolution. I will elaborate the concept of STED microscopy later. The other group that includes techniques like PALM (photo-activated localization microscopy) and STORM (stochastic optical reconstruction microscopy) is called stochastic super-resolution. These techniques are mainly based on random excitation of fluorophores that are not close to each other, thus avoiding the diffraction barrier.

STED Microscopy

When a laser light of particular wavelength excites a fluorophore molecule, the excitation spot becomes bigger due to diffraction of light (figure 44A, left two panels). STED microscopy uses the approach of limiting the diffracted spot size by quenching the fluorescence around the central spot with the help of STED laser to generate the donut shape de-excitation spot (figure 44A central and right two panels). In normal fluorescence, the excited electrons in the atom of fluorophores jump to higher energy levels-S1, S2 and then relax to ground state-G by releasing another photon called fluorescence (figure 44 B left panel). Fluorescence is a spontaneous process and occurs in all directions and thus we can collect it from any side of the fluorophore. In STED microscopy, a second laser that matches the emission wavelength of the fluorophore, the red-shifted STED laser, is applied to quench fluorescence around the central excitation spot. Under the influence of this external electromagnetic radiation, the excited fluorophore jumps to ground state before fluorescence to happen by releasing the second photon of same phase and polarity as the STED laser (figure 42B right panel). As the emitted photon is of the same polarity as the STED laser, it is neglected as excitation light. This is how the STED microscopy de-excites fluorophores and generates smaller fluorescence point to limit diffraction. The smaller the excitation spot the better is the limit of resolution. This technique has been reported to achieve a limit of resolution up to 20-50 nm or even less. Stefan W Hell invented this Nobel Prize winning STED microscopy technique.

The concept of stimulated emission is not new; rather it is applied in the production of LASER light also as you can see that it is the acronym for **L**ight **E**mission by **S**timulated **E**mission of **R**adiation. Stimulated emission in LASER emitter produces photons of same phase and polarity (direction) which results in the monochromatic, collimated and amplified light that travel in one direction for very long distance.

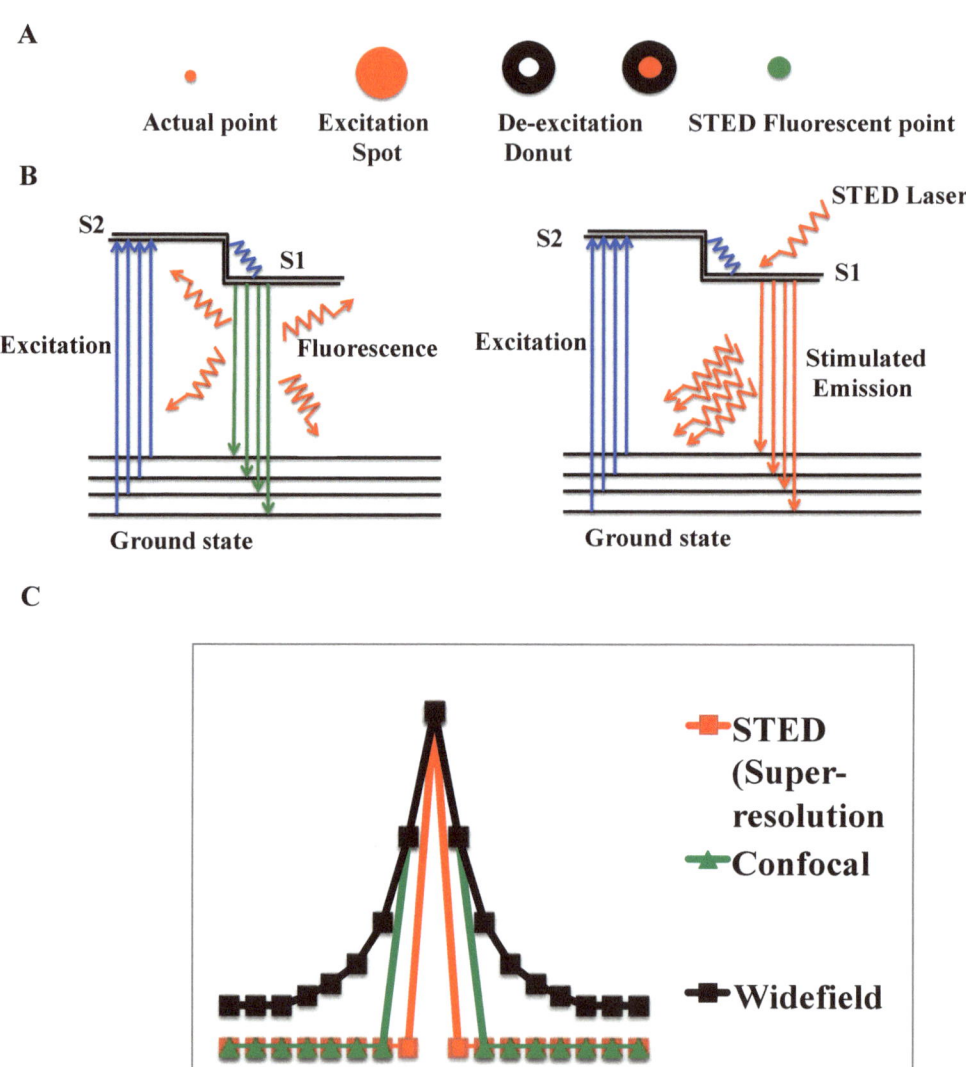

Figure 44

PALM microscopy

The discovery of photo-activable proteins (like PA-GFP) and photo-switchable dyes (like Cy3) has made it possible to achieve super-resolution image by selectively switching on or off a well-defined set of fluorophores. While PALM microscopy uses photo-activable protein the STORM microscopy uses photo-switchable dyes. Normally, the fluorescence spread of nearby fluorophores obstructs the limit of resolution because of diffraction (figure 45A). PALM overcomes this hurdle by selectively switching on few molecules, which are far apart, thus sequentially collecting fluorescence from non-overlapping molecules (figure 45B). The fact that photo-activable proteins can be switched on by a specific wavelength of light and photobleached by a certain wavelength makes it convenient to control the fluorescence of the selected set of fluorophores. Next, we can localize the center of each molecule mathematically by utilizing the point spread function, the number of photons collected, background noise and the pixel size of the camera (figure 45C and D). This approach enables us to track the exact center of fluorescent molecules and undo the diffraction barrier. In this way, PALM microscopy can achieve the limit of resolution as good as ~10 nm.

Related super-resolution technique STORM utilizes the similar approach of exciting distant non-overlapping fluorophores. The discovery of photo-switchable dyes like Cy3-Cy5 that can be switched on or off in certain chemical buffers made it possible to collect fluorescence from a large number of fluorophores in a stochastic manner.

Photoactivated Localization Microscopy (PALM)

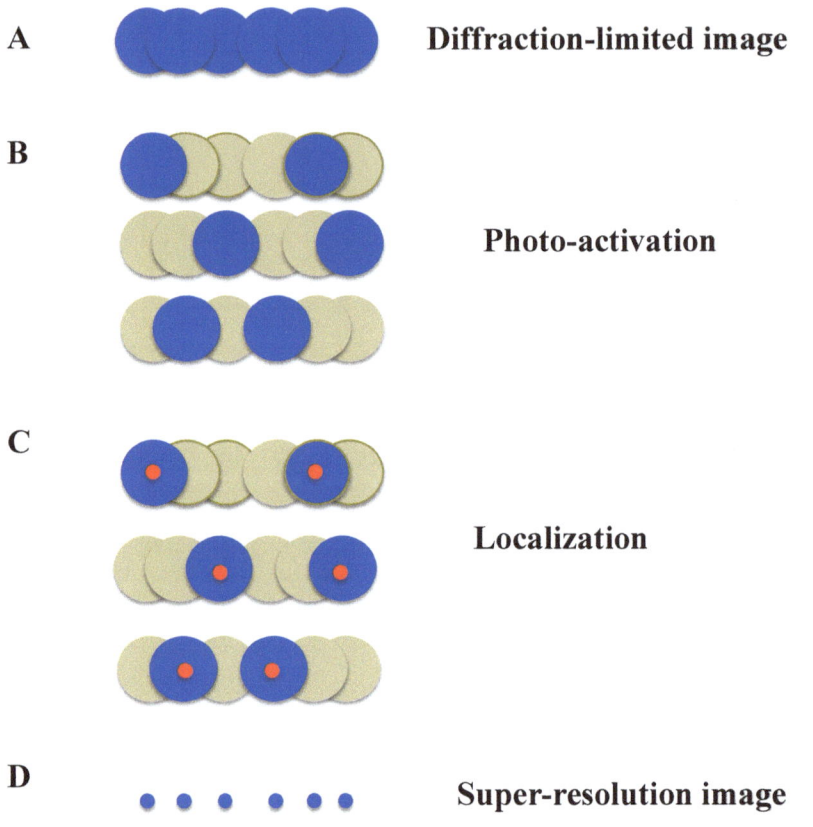

Figure 45

"Eyes never sleep. The electronics in the eye is always on, even in the dark."

Electronics in the eye

The formation of the image of an object on the retina of the eye does not make much sense to vision. It is the electrical signal generated by light (from object) falls on the retina makes vision possible. We can compare the image processing in the eye to image processing in the digital camera. Here, the understanding of light as a particle will help us explain the image sensing and processing. When an image is formed either on the retina of the eye or the photocathode of camera chips or the photomultiplier tube (PMT) in some microscopes (Figure 46A), the image is sensed as an organized mass of photons. When a photon hits the photocathode, it knocks out an electron, which is amplified as electronic current (figure 46B). The electronic current is subsequently converted into the digital signal by the analog to digital converter (ADC) whose sampling rate from pixel to pixel can affect the quality of the digital image. The digital signal is finally projected into the screen of the computer to produce the final image. A similar phenomenon is employed in the classical film camera. A brief exposure to light (coming from the object) causes the release of an electron from halide ions to silver ions in the photographic film. This leads to conversion of the silver ion into metallic silver that produces the latent image in the film, which is further developed into the final picture.

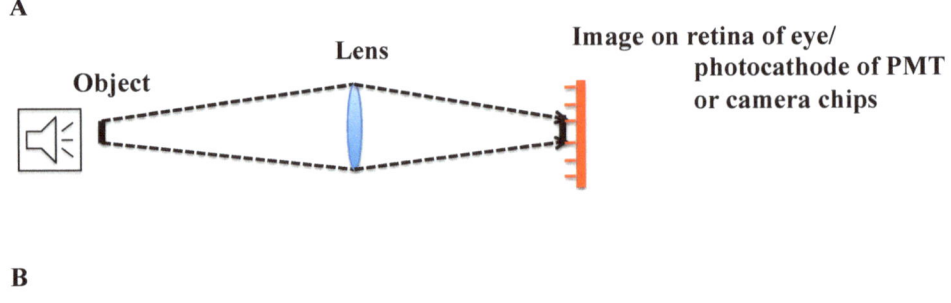

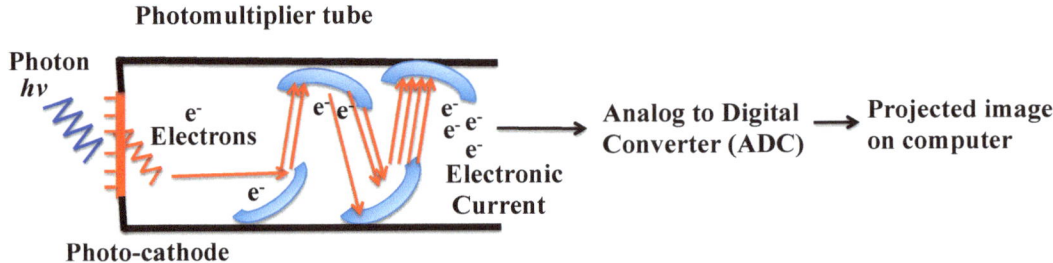

Figure 46

Similar to the digital image processing, the image in our eye is processed through a series of steps in order to generate an electronic signal, which is basically the flow of currents (ions) in optic nerve that connects the eye to the brain. The electrical signal that carries the visual image from the eye to the brain is actually carried by a special type of neuron cell called the retinal ganglion cells (RGC) as shown in figure 47. Let me briefly describe how light is processed in the eye. As we have discussed previously, a real inverted image is produced in the retina of the eye. An image is nothing but light. The photoreceptor cells in the retina perceive this light. There are two main types of photoreceptor cells- the Rods and Cones. These cells are named based on their shape- rod cells have a rod like disc and cones have a conical disc (figure 47). Rods are responsible for bright or dim light vision whereas cones are for color vision. These cells are concentrated in the region of fovea but are absent in the region called blind spot where the optic nerve leaves the eye.

A special pigment in these photoreceptor cells called Opsin (Rhodopsin in Rods or Conopsin in Cone cells) is the actual receiver of light signal. A different kind of light sensitive pigment called Melanopsin in photosensitive retinal ganglion cells (pRGC) is responsible for circadian rhythm. Circadian rhythm is the 24hr biological clock that our body adjusts to day and night. The fact that we wake up in the morning basically involves a process where sunlight is perceived by the pRGC cells, which convey the message to the suprachiasmatic nucleus (SCN) of the brain to regulate the circadian rhythm.

Opsins are distributed heavily in the outer disc of rod cells just like the antenna of a TV (figure 47). The actual molecule (associated with opsin) that absorbs photon is called 11-Cis-retinal, a derivative of vitamin A. The absorption of light leads to a conformational change of cis-retinal to all-trans-retinal. This causes Opsin to activate another protein, Transducin, which in turn activates the enzyme Phospho-Di-esterase that breaks down c-GMP. Reduction in the level of c-GMP shuts down sodium channel resulting in the hyperpolarization of rod cell. This leads to closure of voltage-gated calcium channel and thus inhibits the release of neurotransmitter "Glutamate". Drop in the level of Glutamate triggers a set of on center bipolar cells. These bipolar cells then send a signal through the release of neurotransmitter glutamate to retinal ganglion cells, which generate current-- called "action potential" by the influx of ions (mainly sodium) through a series of ion channels. RGC cells convey the visual signal in the form of action potential through their long axons to the visual cortex of the brain. A nerve signal is basically the flow of ions, unlike the digital camera where the current is basically a flow of electrons. The work by British scientists Hodgkin and Huxley on the axons of the giant squid in the Marine Biology Laboratory (at Plymouth, UK) led to the understanding of action potential and visual signal processing. They were awarded the Nobel Prize for their groundbreaking work in 1963. On a trip for microscopy coursework at Plymouth, UK, in 2015 I got a chance to visit the original lab of Hodgkin and Huxley.

To sum it up, the light is converted into an electronic signal in the eye and conveyed to the visual cortex of the brain for visual perception. Regardless of whether we open or close our eye, awake or asleep, firing of optic nerves continues some way or the other and that is how we do see imaginary pictures in dreams too.

At the beginning of the book, I had mentioned about the allegory of the cave. Let me try to explain it scientifically. When the slave suddenly came out from the darkness of the cave to see the reality in front of the fire, the brightness of light, upon sudden exposure, blinded him. We often come across such situation if we switch on a light while sleeping. This is because of unavailability of Opsins, the active form of retinal, c-GMP and other molecules as the photoreceptor cells in the dark are tuned to the low level of basal signaling in dark. As we slowly adapt to light (from darkness), these molecules are regenerated and transported to their proper location in order to be in sufficient number to receive light. There is an on-going cyclic process of photoreceptor regeneration and use in the eye.

$$\text{Rhodopsin} \longleftrightarrow \text{Retinal + Opsin.}$$

When the slave slowly adapted to light and gained vision outside the cave, his eyes were actually undergoing the course of recovery of active photoreceptors. Finally, when the slave returned to the cave, the sudden exposure to the darkness of the cave blinds him again. We come across a similar situation when we enter a dark movie theatre and struggle to find our allocated seats. As we wait a few minutes in the dark we gain dim light vision and become comfortable to see inside the dark theatre. This is because, in bright light, the photoreceptors are saturated or bleached, so no active rhodopsin molecules are available for vision in dim light in the beginning as we enter the dark room. At the early time of exposure to the dark, the cones take over, but gradually the rods take control of the dim light vision. These processes take a few minutes.

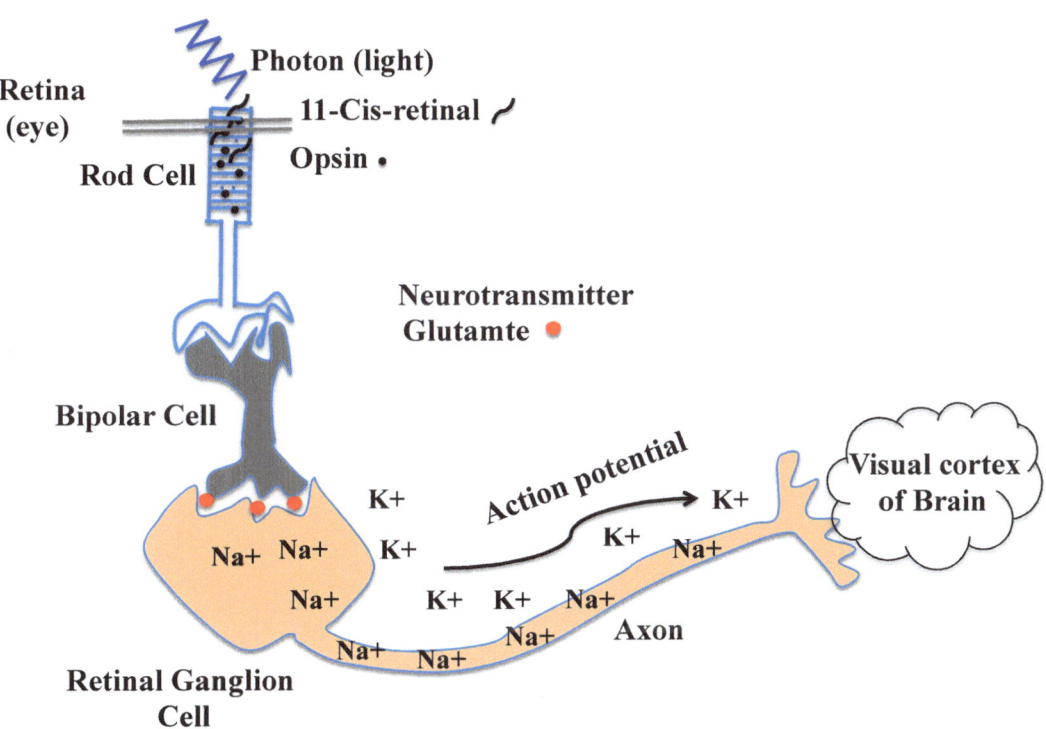

Figure 47

Conclusion

      Thousands of years have passed since human beings have been curious to understand the exact nature of light. It might still be an open question for many generations to come. Although we cannot see what the elementary particle of light- "photon" looks like, various phenomena associated with the behavior of light have helped us understand what light is. At least in any particular setting, we can imagine how light behaves. In the first part of the book, I tried to explain some of the fundamental properties of light. The greatest gift of the sun to the earth "**light**" has helped the human being to see the real mystery of nature beyond the natural power of the eye. It helps us (human being) to see micro or nanometer size organisms, which no other animals in the world can see, thanks to the invention of the microscope. However, as light helps us to see, it also hinders us from seeing beyond a limit because of the diffraction barrier. Credit goes to the development of super-resolution techniques, which have helped us to see beyond the diffraction barrier. So far, a resolution limit of as small as ~10 nm has been achieved by applying super-resolution light microscopy techniques. A protein can be of 1 nm or even smaller in size. The typical "coke can" like structure of GFP protein is ~2.4 nm (as determined by X-ray-crystallography). Other molecules inside the cells are even way smaller than a nanometer (of pm size). A lot of work needs to be done to further improve the limit of resolution in light microscopy. Non-conventional microscopes like the electron microscope have the capacity to generate image of even thousand-fold better resolution (up to ~50 pm) than a super-resolution light microscope can do. However, it is based on entirely different concept than light and optics. As I mentioned before the electron beam can also be subjected to diffraction. Nevertheless, electron microscope has other limitations and cannot be used as a conventional microscope. My curiosity is to know what is the smallest molecule or structure that human being can see with the help of microscope that uses "light" which is primarily designed by mother nature for vision. The author hopes that scientists will soon come up with techniques that will help undo the diffraction of light and see the structures of sub-nanometer or picometer scale with light microscopy.

Glossary

Amplitude- The distance between the central line to the either the top of trough or bottom of the crest.

Black body- An idealized body that absorbs all incident electromagnetic radiation.

Bright-field microscopy- A microscopy approach where field looks bright and the object looks darker.

Compound microscope- A microscope with at least two lenses.

Condenser lens- The lens that focuses light on the sample.

Confocal microscope- A type of microscopy that uses pinhole apertures to obstruct out of focus light.

Convex lens- A lens with both the surfaces bulging in the middle.

Electron- The particle with negative electric charge orbiting the nucleus of an atom.

Diffraction- The process by which waves spread out after passing through a narrow aperture.

Dispersion- The phenomenon by which the constituent colors of light are separated after passing through a prism.

Electromagnetic wave- A wave with both oscillating electric and magnetic field.

Eyepiece- The lens closer to eye in the microscope.

Evanescent wave- The near-field fading electromagnetic wave.

Fluorescence- The process of release of higher wavelength light after absorption of lower wavelength light by molecules.

Frequency- The number of occurrence of events per unit time.

Interference- A phenomenon where two waves superpose to yield a third wave of higher or lower amplitude.

GFP- A Green Fluorescent Protein discovered from jellyfish *Aequorea.*

GSD- Ground State Depletion

IRM- Interference Reflection Fluorescence Microscopy.

LASER- Light Amplification by Stimulated Emission of Radiation.

Limit of resolution- The minimum resolvable distance between two adjacent points.

Magnification- Increase in the size of the image caused by the lens.

Microscope- An instrument that that helps us to see objects smaller than the power of the naked human eye.

Numerical Aperture- The product of refractive index and sine of semi-angle of the cone of light received by the lens.

Objective lens- The lens closer to the object in the microscope.

Oscillation- Repetitive vibration around central value.

PALM-Photo Activated Localization Microscopy.

Photomultiplier tube (PMT)- A phototube that amplifies current produced by an incident photon.

Photoelectric effect- The phenomenon of knocking out an electron from the atom by an incoming photon.

Photon- The elementary particle of light.

Pinhole camera- A camera with a pinhole without a lens.

Polarization- The direction of oscillation of a wave.

Quantum- The smallest unit of electro-magnetic energy.

Reflection- Change in direction of wave at the interface between two media where the wave returns to the same medium from where it arrived.

Refraction- Change in direction of electromagnetic wave after passing obliquely from first to second media of different density.

Refractive Index- The ratio of the speed of light in the vacuum to that of the medium.

Signal transduction- The process by which signal from the receptor is transmitted through a series of molecules.

Simple microscope- A microscope with only one lens.

Sine wave- A periodically oscillating wave.

STED- Stimulated Emission Depletion microscopy.

STORM- Stochastic Optical Reconstruction Microscopy.

Super-resolution – Resolution achieved beyond the diffraction barrier.

TIRF- Total Internal Reflection Fluorescence.

Wave- Oscillation accompanied by the transfer of energy.

Wavelength- The distance between two adjacent crests or troughs of a wave.

Wide-field microscopy- A microscopy technique where light from the whole field of view is collected at the same time to generate the image.

4Pi Microscope- The microscope that has two objective lenses to simultaneously collect light from the sample.

Index

Amplitude, 21, 24, 25

Blackbody, 8,

Bright-field microscopy, 61

Compound microscope, 47-48, 73

Condenser lens, 47-48, 73, 76

Cones, 39, 84-85, 87

Confocal microscope, 64-65, 73, 76

Convex lens, 13, 40-43, 51, 55-56

Electron, 32-34, 62, 79, 83

Diffraction, 23, 26, 34, 69, 71, 73

Dispersion, 12, 13

Electromagnetic wave, 18-20

Eyepiece, 47-48

Evanescent wave, 66-67

Fluorescence, 62-63, 79

Frequency, 19-20, 31-33

Interference, 23-26, 67-68

GFP, 63, 88

GSD, 78

IRM, 67-68

LASER, 26, 28, 73, 74, 79

Limit of resolution, 76-81

Magnification, 49, 50

Microscope, 40, 45-48

Numerical Aperture, 59, 76

Objective lens, 47-50, 57-58

Oscillation, 21

PALM, 81-82

Photomultiplier tube (PMT), 83-84

Photoelectric effect, 31-32

Photoreceptor, 7, 84-87

Photon, 23, 31-33

Pinhole camera, 36-37

Polarization, 27

Quantum, 31, 62

Reflection, 30-31, 66-67

Refraction, 22-23, 41-42

Retina, 38, 84-87

Rods, 39, 84, 85, 87

Refractive Index, 22, 40, 66-67

Simple microscope, 47

Sine wave, 21-22

STED, 78-80

STORM, 78, 81

Super-resolution, 69, 78, 80

TIRF, 66

Wave, 13, 17-22

Wavelength, 13, 15, 19-22

Wide-field microscopy, 64

4Pi Microscope, 58

www.ingramcontent.com/pod-product-compliance
Lightning Source LLC
Chambersburg PA
CBHW051155220526
45473CB00003B/786